PLANT FAMILIES
How To Know Them

Pictured-Keys for determining the families of nearly all of the members of the entire Plant Kingdom.

Second Edition

H. E. JAQUES

Professor of Biology
Iowa Wesleyan College

WM. C. BROWN COMPANY PUBLISHERS
Dubuque, Iowa

Library of Congress Catalog Card Number: 48-4171
ISBN 0—697—04841—1 (Cloth)
ISBN 0—697—04840—3 (Paper)

Twenty-sixth Printing, 1984

THE PICTURED-KEY NATURE SERIES

How To Know The—

AQUATIC PLANTS, Prescott, 1969
BEETLES, Jaques, 1951
BUTTERFLIES, Ehrlich, 1961
CACTI, Dawson, 1963
EASTERN LAND SNAILS, Burch, 1962
ECONOMIC PLANTS, Jaques, 1948, 1958
FALL FLOWERS, Cuthbert, 1948
FRESHWATER ALGAE, Prescott, 1954, 1970
FRESHWATER FISHES, Eddy, 1957, 1969
GRASSES, Pohl, 1953, 1968
GRASSHOPPERS, Helfer, 1963, 1972
IMMATURE INSECTS, Chu, 1949
INSECTS, Jaques, 1947
LAND BIRDS, Jaques, 1947
LICHENS, Hale, 1969
LIVING THINGS, Jaques, 1946
MAMMALS, Booth, 1949, 1970
MARINE ISOPOD CRUSTACEANS, Schultz, 1969
MOSSES AND LIVERWORTS, Conard, 1944, 1956
NON-GILLED FLESHY FUNGI, Smith-Smith, 1973
PLANT FAMILIES, Jaques, 1948
POLLEN AND SPORES, Kapp, 1969
PROTOZOA, Jahn, 1949
ROCKS AND MINERALS, Helfer, 1970
SEAWEEDS, Dawson, 1956
SPIDERS, Kaston, 1953, 1972
SPRING FLOWERS, Cuthbert, 1943, 1949
TAPEWORMS, Schmidt, 1970
TREMATODES, Schell, 1970
TREES, Miller-Jaques, 1946, 1972
WATER BIRDS, Jaques-Ollivier, 1960
WEEDS, Wilkinson-Jaques, 1959, 1972
WESTERN, TREES, Baerg, 1955, 1973

Printed in the United States of America

INTRODUCTION

WHEN life and living things are considered, nothing else is so basic as the plants. All other life is directly or indirectly dependent upon the green plants for sustenance. An intimate knowledge of plants is necessary for many vocational fields. Culture demands some knowledge of plants. Everyone may find interest and relaxation in studying or associating with them.

The total number of plants known to science is so great that a lifetime would scarcely suffice for one to learn to recognize all of them. The family, which is an aggregation of similar plants, makes a unit highly important for study. The number of families in the plant kingdom is sufficiently small that with a reasonable amount of study one can place most of the plants he sees anywhere in their proper relationship and understand them much better. In our judgment, no other division offers as good possibilities for a broad general understanding of plants and animals as the family group. It is for this reason that this book has been written.

The Pictured-Key Nature Books are not based on new research. The effort instead has been to take the important facts of plants and animals and to make them more understandable for the beginner and all students who appreciate clarity. In "Plant Families" all groups of plants have been considered and one or more plants used as examples of each family. Species of plants commonly seen have been chosen wherever possible for these examples. Some families of small consequence have been purposely omitted in the interests of making the keys easier to handle. With such a large field to cover, space has not permitted much to be said about any one plant. When the student finds himself especially interested in some particular group of plants, he will need to refer to one of the many excellent books on the group of his choice.

Many good friends have helped us. Professor Henry S. Conard of Grinnell College, whose studies of Mosses are well known, has written the keys for the entire division Bryophyta. Arlene Knies, Mabel Jaques Cuthbert, Francesca Jaques Stoner and our good neighbor, Betty Laird Swafford have made most of the drawings. It's grand to have faithful friends; we want to thank them all.

In this revised edition a good number of changes have been made to the illustrations and text, and the entire book set in new format.
Mt. Pleasant, Iowa,
March 1, 1948.

CONTENTS

SOME PLANT FACTS

THE KING was in his counting house, counting out his money." Kings have not always given their first attention to the welfare of their subjects.

Man is "king" of all the animals. He writes the books and says so. But there are no plants which publish claims to authority, so it may be a question just which one heads up the Plant Kingdom.

That doesn't matter anyway for his subjects really count for more than a king. In this case there are 250,000 different kinds of subjects, but no one has ever tried to even guess how many individuals.

We took a census once of just the trees growing in Mt. Pleasant, —an unusually nice little town. They added up 15,998 individual trees. We would have been crazy, of course, to have attempted to count the dandelions or the grass plants, or the bacteria.

So much for random thoughts. Since there are some important things that should be said about plants, we'll get on to a few of them.

PLANT REQUIREMENTS

Moisture, an acceptable temperature, and for most plants, soil and sunlight are necessary for their growth. Where all of these conditions prevail in highly satisfactory degrees, plants are most abundant and at their best. In moist tropical areas plants if left to themselves grow into the nearly impenetrable jungle. Once the jungle growth is removed and such an area set with useful plants, the yield may be prodigious.

Temperate regions for part of the year are too cold or may have such limited rainfall as to support only a fair plant coverage. Deserts have everything else, yet grow but little for want of moisture. The seas have the moisture and usually the food materials and sufficient warmth. In their upper strata the water courses often produce a prolific plant growth but fail at greater depths through being unable to meet the light requirement. As the poles are approached, temperature becomes the limiting factor and plant life while present, is greatly reduced in size and in numbers of species and of individuals.

SOME EARLY PLANTS

Plants are age-old. Even at the dawn of the Cambrian period — some would say nearly a billion years ago — many of the simpler plants had made a good start. Ferns have been on the way perhaps half a billion years while flowering plants have likely been beautifying the world with their blossoms and fragrance for more than fifty million years. Plant styles have undergone many changes through these long ages and numerous species have flourished for a time and then become extinct. Abundant fossil remains reveal vegetation growths that in some ways greatly out-stripped the best we know today.

1

PLANT FREEDOM

The green plants live an independent life. With the sunlight for a motive force, they combine the carbon dioxide of the air with water to form much of their tissues. With a few more elements taken from the soil, they are self sufficient.

Other great groups of plants have been "on relief" so long as to lose their chance of independence. Now with their chlorophyll gone, they must live as parasites or saphophytes and like all the animals, are directly or indirectly dependent upon the green plants for food. The green plants then become king of all living things.

LARGE AND SMALL PLANTS

Plants on the whole display a wide range of sizes. Some bacteria are so small that more than six thousand billion would be required to fill a cubic inch of space. At the other extreme, vines occasionally exceed a length of one thousand feet. Now and then a tree may be over 300 feet high with a trunk more than 30 feet in diameter. One tiny bacterium would have about the same size ratio

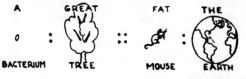

to the big tree as a well fed house-mouse would have to the entire earth and all that is in it.

PLANT PROJECTS

The best way to know and understand plants is to live with them. For the student, some collecting or research project if thoughtfully pursued is sure to pay well in pleasure and knowledge. "How to Know the Trees" suggests 26 tree study projects (p. 12) while in "Living Things—How to Know Them" a chapter is entitled "More than 100 Suggestions for Nature Study Projects" (p. 3 - Rev. Ed.). These offer many good suggestions to which the ingenious teacher or student can readily add other helpful ones.

COMMON VS. SCIENTIFIC NAMES

Plants that have common everyday uses or relationships are known by "common names". In fact, many of them have several common names and that is how the trouble begins. For instance, *Abutilon theophrasti*, a widely distributed and all too abundant Asiatic weed, is referred to by the following, Velvet Leaf, Indian Mallow, Butterprint, Button-weed, Pie-print, Mormon-weed, Cotton-weed, Indian Hemp, Sheep-weed, American Jute, Pie Marker, etc. If a plant is cosmopolitan each language may also have one or more names for it.

To make the confusion still worse the same name is frequently applied to several different plants, leaving the hearer in doubt as to which one is meant.

To ease this difficulty, Linnaeus, a Swedish naturalist around 1760, devised a system of scientific names which would be world wide in their application. He did such a good job that his scheme is still in use, and while it is not perfect the plan of having one universal scientific name for each plant has many advantages over common names.

Some violets are yellow, some purple, some white; some are large, others are small; the leaves are sometimes deeply cut and some are not cut at all. There are so many differences among the

violets that many kinds ("species" is a better word) are recognized. All of these are enough alike in some essential characters that they are unmistakably related. Each one shows its relationship to its group. Such an aggregation is known as a genus and is given in this case the name *Viola* which is the Latin word for violet. Just as the Smiths have their John, William and Mary, so there are *Viola odorata, Viola pubescens, Viola pedata, Viola tricolor*, and many others. These scientific names are made up of a Latin noun the genus. It is always capitalized. The word following is the species and is a Latin adjective modifying the generic noun or a noun in apposition with it. The species name often reveals some important characters of the plant. Thus the four voilets named above are respectively, very fragrant, have hairy stems, have leaves resembling a bird's foot, or are variously colored (pansy). The word or abbreviation following the species is the "authority" or "author" and tells what scientist proposed this scientific name. Zoologists have ruled that all species names begin with a small letter; some botanists prefer to follow that same plan. Scientific names are printed in Italic type, or underscored when written or when Italic type is not available.

A few basic plant facts have been quickly mentioned. There is much to be known about plants and many excellent books to tell it. The reader whether a beginner or one who has long loved plants is wished a continuing of happy experiences with these, our faithful friends.

SOME HELPFUL BOOKS

 small book like Plant Families, at best, can present only a general view of the entire plant kingdom. If one becomes especially interested in any particular group of plants he will need to refer to books that specialize in his chosen field. Some highly useful ones are suggested below, but this list must not be thought of as being at all complete. The attempt has been only to include manuals useful in identifying plants.

ALGAE

"The Fresh-water Algae of the United States",G. M. Smith

"The Green Algae of North America", F. S. Collins

"Cryptogamic Botany", Vol. 1 Algae and Fungi, Smith

"Algae, the Grass of Many Waters", Tiffany

FUNGI

"Bacteriology", Tanner

"The Fungi Which Cause Plant Disease", Stevens

"Agaricaceae of Michigan", Kauffman

"Mushrooms, Edible, Poisonous, etc.", Atkinson

"Manual of the Rusts in the United States and Canada", Arthur

"Tne Biology of Bacteria", Henrici

"The Lichen Flora of the United States", Fink

"The Myxomycetes", Macbride & Martin

"Manual of Vegetable Garden Diseases", Chupp

"Elements of Plant Pathology", Melhus & Kent

"Introduction to Plant Pathology", Heald

"Principles of Plant Pathology", Owens

"Filterable Viruses", Rivers

"One Thousand American Fungi", Chas. McIlvaine

"Mushrooms and Toadstools", Gussow & Odell

"The Mushroom Book", N. L. Marshall

MOSSES AND LIVERWORTS

"Mosses with a Hand-lens", A. J. Grout

"Mosses with Hand-lens and Microscope", A. J. Grout

"Hepaticae of North America", Frye & Clark

"How to Know the Mosses", H. S. Conard

4

FERNS

"Ferns of North Carolina", Blomquist

"Ferns of the Northwest", Frye

"Ferns of Tropical Florida", Small

"Guide to Eastern Ferns", Wherry

FLOWERING PLANTS

"Wild Flowers", House

"Hortus", Bailey

"Flora of the Prairie and Plains of Central North America", Rydberg

"Wild Flowers Worth Knowing", Blanchan

"How to Know the Spring Flowers", Mabel Jaques Cuthbert

"Manual of the Grasses of the United States", Hitchcock

"Illustrated Flora of the Northern United States and Canada", Britton & Brown

"How to Know the Trees", H. E. Jaques

"Illustrated Flora of the Pacific States", Abrams

"Manual of Botany", 7th Edition, Asa Gray

"Field Book of American Wild Flowers", Schuyler F. Mathews

"Manual of Cultivated Plants", L. H. Bailey

"Flora of the Southeastern United States", Small

"Flora of the Rocky Mountains", P. A. Rydberg

"Field Book of Western Wild Flowers", Armstrong

"Wild Flowers of California", M. E. Parsons

"Plants of Iowa", H. S. Conard

"How to Know the Fall Flowers", M. J. Cuthbert

"Key to Some Common Weeds", Paul B. Mann

"Field Book of American Trees and Shrubs", Schuyler F. Mathews

"Handbook of the Trees of the Eastern United States", R. B. Hough

"Trees and Shrubs of the Rocky Mountains", Burton O. Longyear

THE FAMILY PICNIC

HOW TO USE THE KEYS

The use of keys for identifying plants and animals dates back many years. The addition of pictures to supplement the keys is more recent and makes their meaning clearer. The use of keys is much like traveling a strange region where the road intersections are well marked. If the traveller reads the signs intelligently and follows their instructions, he should have no trouble. A little understanding of terms and the use of care in selecting the right direction each time should find the correct family to which the plant belongs.

It will be noted that the key statements are set in opposing pairs which are numbered alike but lettered differently. We find a toadstool in our front yard and wish to know to what family it belongs. With specimen at hand we start at page 7 and compare 1a with 1b. Noting that our specimen has no flowers,' true roots, or leaves, we try "Division Thallophyta, page 8".

This time we find three statements to be compared (1a, 1b, and 1c); 1b fitting the case shows us our specimen belongs to the Fungi on page 44 and we compare 1a and 1b to select 1b and go as directed to number 2 where 2b is seen to be right and we then consider 21 to select 21b and then 22b, 60b, 64b, 65b, 67b successively. Finally 69b reveals that our specimen belongs to the "Family Agaricaceae". Now if we will go back through the keys noting the facts that lead us to our decision, we will have a good technical description of the Agaric mushrooms.

Near our toadstool may grow a dandelion and we wonder about its family. Beginning again at page 7, 1b, 2b, and 3b send us to the seed bearing plants, page 89, where it is found to be an Angiosperm (1b) and a Dicotyledon (4b). On page 100 we select 1b then go in turn to 25b, 105b, and 106a to learn that the pesky dandelion is associated with the asters, daisies, and sunflowers in the great family Compositae.

In the back of the book we find that a list of the families of plants arranged according to their relationship begins at page 155. If each plant studied is checked in this list and its relatives noted, the student will get a better concept of the plant kingdom, as well as having a graphic record of his progress.

PICTURED-KEYS FOR IDENTIFYING THE FAMILIES OF THE ENTIRE PLANT KINGDOM

There are more than 250,000 known species of plants. A life time of specialization would be needed to know most of them intimately. But these many species fall into a few hundred families. An earnest

student or careful observer can acquaint himself with so many of these families that he will recognize the relationships of the majority of the plants he sees, no matter how widely he travels. The family makes the best unit for a thorough general knowledge of plants or animals.

To secure a clear picture of the plant kingdom is is necessary to begin with the larger groups. The reader will now gather up his equipment of keen eyes, alert mind, and good perseverance and he is ready to start on a most fascinating journey through the Kingdom of Plants.

KEY TO THE PRINCIPAL GROUPS OF PLANTS

The plants are usually divided into four great groups. It is commonly recognized that some of these groups are "artificial". The Thallophyta in particular includes many plants which are somewhat alike in their simplicity but which are evidently widely disrelated.

1a Plants without distinction of root, stem and leaf; without flowers, both aquatic and terrestrial; algae; fungi, and lichens. *"HAVE NO ARCHEGONIA"*. Figs. 1 to 200.   Division **THALLOPHYTA** page 8

1b Plants with distinct leaves; with or without roots or flowers. 2

2a Small plants (to 4 or 5 inches tall) with green or gray-green leaves, or tiny leaf-like forms on damp earth or floating on water. No true roots or flowers. (Some small oval green discs floating on quiet water with roots suspended beneath are flowering plants and do not belong here.) *"HAVE ARCHEGONIA BUT NO VASCULAR BUNDLES"*. (Mosses and Liverworts.) Figs. 201 to 253.
                    Phylum **BRYOPHYTA** page 69

2b Plants with true roots and vascular bundles; mostly with veiny leaves. 3

3a Plants without flowers or seeds, herbs (not woody), propagated by spores. Figs. 254 to 265. *"HAVE VASCULAR BUNDLES BUT NO SEEDS"*. (The Ferns, etc.)   Phylum **PTERIDOPHYTA** page 86

3b Plants with flowers (sometimes very simple; one tiny stamen or one pistil may constitute a flower) and seeds. *"HAVE SEEDS"*. Figs. 266 to 472. (The seed-bearing plants.)
                    Phylum **SPERMATOPHYTA** page 89

THE THALLOPHYTA

1a (a, b, c) Plants green, with chlorophyll, thus organizing their own food by photosynthesis. Figs. 1 to 116.
Sub-division PHYCOPHYTA, The Algae....2

1b Plants without chlorophyll (not green). Living parasitically on living plants or animals or as saprophytes on dead, organic matter. Figs. 127 to 200.   Sub-division MYCOPHYTA, The Fungi page 44

1c Dual organisms in which a green plant species (alga) is held in parasitic embrace by a species of fungus. Usually gray-green or yellow-green but sometimes displaying bright colors. Common on tree trunks, rocks and often growing directly on the ground. Figs. 117 to 126.
The Lichens page 41

KEYS TO THE FAMILIES OF THE ALGAE

2a Plant cells without recognizable nucleus or plastids. Coloring matter (usually blue-green) diffused throughout the cell. Figs. 1 to 9. (The Blue-Green Algae) 3

2b Not as in 2a. Plant cells with distinct plastids and of various colors and shapes. 10

THE BLUE-GREEN ALGAE (MYXOPHYCEAE)

3a (a, b, c) Unicellular or colonial in habit; never forming filaments. Reproduction vegetative only. Figs. 1 and 2. 4

3b Cells forming definite filaments, but no endospores. Figs. 5 to 9...6

3c Cells usually growing on other plants; solitary, colonial or forming small filaments. Forms endospores. Figs. 3 and 4. 5

ORDER CHROOCOCCALES

4a Cells always in colonies which grow prostrate upon rocks or other bodies and which have erect column like outgrowths. Largely marine.
Family ENTOPHYSALIDACEAE

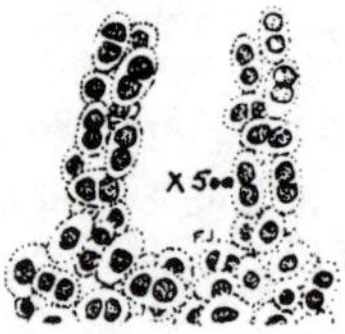

Fig. 1. *Entophysalis magnoliae* Farl.
The family is largely marine though some species grow in fresh water. Rocks between the tide lines form the favorite marine habitat.

Figure 1

4b Cells solitary or in colonies but never with column like outgrowths.
Family CHROOCOCCACEAE

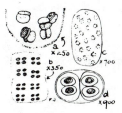

Figure 2

Fig. 2. a, *Chroococcus turgidus* Nag.; b, *Merismopedia puncata* Mey.; c, *Synechococcus aeruginosus* Nag.; d, *Gleocapsa* sp.

Many known species of diverse form make this a large family. The numerous tiny free floating species make an important group in water contamination.

ORDER CHAMAESIPHONALES

5a Thallus multicellular as a result of cell division. In spore formation an entire cell is completely divided into endospores.
Family PLEUROCAPSACEAE

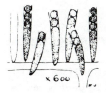

Figure 3

Fig. 3. *Hyella fontana* H. & J. a, vegetative thallus; b, cells forming endospores. There are a few fresh water forms but most of them are marine. For the most part they grow attached to other plants.

5b Cells solitary although often many growing close together. Spores cut off at end of cell one at a time. Family CHAMAESIPHONACEAE

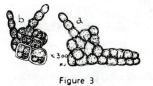

Figure 4

Fig. 4. *Chamaesiphon incrustans* Grun.

Rather common as an epiphyte on larger freshwater algae. Most of the family are marine species.

ORDER HORMOGONALES

6a Thallus a filament of uniform diameter except possibly the terminal cells may be smaller. There are no heterocysts (empty cells). Several filaments often bound together with a sheath.
Family OSCILLATORIACEAE

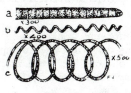

Figure 5

Fig. 5. a, *Oscillatoria* sp; b, *Spirulina major* Kutz.; c, *Lyngbya contorta* Lemm.

This plant is a classic among the "bluegreens". Its "oscillations" consist of moving the tip of its filament in a circle. Many folks do but little better.

9

6b Heterocysts (empty cells) present in the thallus. Filaments often tapering whip like though sometimes of uniform thickness. Figs. 6 to 9. 7

7a Filaments branched. Figs. 8 and 9. 9

7b Filaments unbranched. Figs. 6 and 7. 8

8a Filaments nearly the same diameter throughout; covered with a gelatinous sheath.   Family NOSTOCACEAE

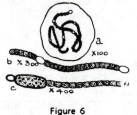

Figure 6

Fig. 6. a, *Nostoc* sp.; b, *Anabaena* sp.; c, *Cylindrospermum* sp.

The pale blue-green gelatinous bean-sized colonies of *Nostoc* are sometimes found in great abundance in clear spring-fed lakes. If they could be kept permanently they would rival pearls for beauty.

8b Filaments gradually tapering to one or both ends.   Family RIVULARIACEAE

Figure 7

Fig. 7. a, *Sacconema rupestre* Borzi.; b, *Rivularia* sp.

Colonies of *Rivularia* are often arranged with the basal ends of the filaments standing at the center of a sphere and the whiplike ends radiating to the circumference. Such balls are frequently 2 or 3 mm. in diameter.

9a Filaments with false branches; enclosed in a sheath.   Family SCYTONEMATACEAE

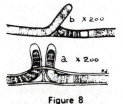

Figure 8

Fig. 8. a, *Scytonema* sp.; b, *Tolypatrix* sp.

False branching is characteristic of several groups of algae. It's something like propaganda or stealing 2nd base;- you must look close to see exactly what goes on.

9b Filaments with true branches; sheathed; often more than one cell in diameter.   Family STIGONEMATACEAE

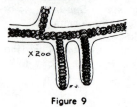

Figure 9

Fig. 9. *Stigonema turfaceum* Cooke.

The members of this genus grow in damp soil or in fresh water and are fairly common. Filaments may rebranch again and again or may be wholly unbranched.

10a Algae that are bright "grass-green" in color. Figs. 10 to 52. 11

10b Algae with red, brown or yellow pigments more or less masking the green chlorophyll. Figs. 53 to 116. . . . .**48**

11a Plant fine; relatively small. Figs. 11 to 52. 12

11b Plant coarse, at least a few inches long, branched, with nodes from which arise whorls of cylindrical leaf-like parts. These in turn bear whirls of still smaller branches. Often incrusted with lime.   Order **CHARALES**
Family **CHARACEAE**

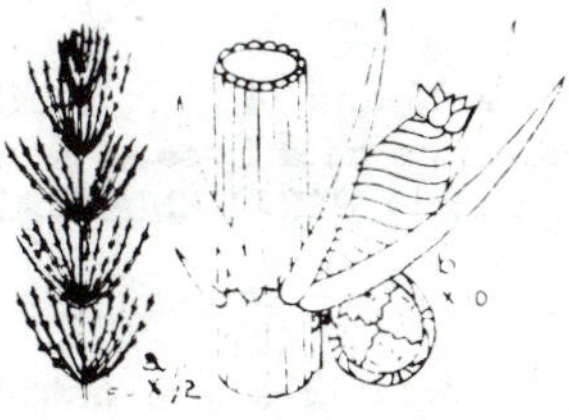

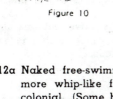

Figure 10

Fig. 10. *Chara fragalis* Derv. a, branch of plant; b, reproductive organs.

This is the only family of its order and class and has comparatively few species though they are fairly common and widely distributed. These plants are so different from all other green algae that they have been given a phylum of their own. These plants attain a height of several inches and are quite complex in their structure. Reproduction is always sexual. *Nitella* is another common genus.

————————*————————

12a Naked free-swimming animal like cells with one, two or rarely more whip-like flagella at the anterior end. Some species are colonial. (Some have a non-motile encysted stage.) Sexual reproduction unknown. Figs. 11 to 14. 13

12b Microscopic single celled plants to large thallus-like plants. Cells with walls of cellulose. The flagella of motile forms usually number 2 or 4, and are equal in length. Most species of the phylum have sexual reproduction. Figs. 15 to 52. 16

ORDER EUGLENALES

13a Cells non-motile and living attached to crustaceans and rotifers in indefinite masses or branching colonies.

Family **COLACICEAE**

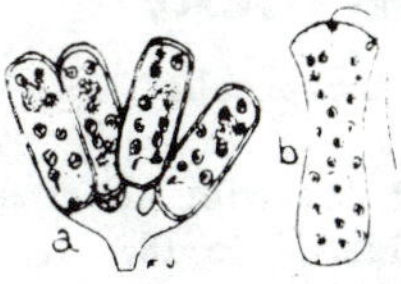

Figure 11

Fig. 11. *Colacium calvum* Stein. a, colony; b, rather rare temporary motile stage.

This family is sometimes assigned to a separate order. And some college students think they invented hitch-hiking. Even very simple plants like these have been at it for millions of years.

13b Vegetative cells unattached and swimming by means of one or more flagella. Figs. 12 to 14. 14

14a Cells usually with green chloroplasts; if colorless an eye-spot is always present, the same as in the colored species. Sexual reproduction is unknown. **Family EUGLENACEAE**

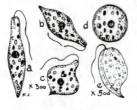

Figure 12

Fig. 12. a, b, c, *Euglena viridis,*-changing forms of active cell; d, encysted cell which will live through drought or low temperatures; e, *Phacus acuminatus* Stokes.

These highly interesting forms often become exceedingly abundant. They may be suspected by the very bright shade of green covering the pool of water.

14b Cells colorless and having no eyespot. Figs. 13 and 14. 15
15a Cell with a pharyngeal rod (elongated structure near base of flagellum). **Family PERANEMACEAE**

Figure 13

Fig. 13. a, *Urceolus cyclostomus* Mer.; b, *Entosiphon sulcatum* Stein.

Some species of the family have but one flagellum; others have two in which case one extends ahead and the other trails.

15b Cells without a pharyngeal rod. **Family ASTASIACEAE**

Figure 14

Fig. 14. *Astasia dangeardii* Lemm.; b, *Menoidium incurvum* Kleb.

Flagellated animals of this type usually pull themselves through the water instead of carrying the flagellum behind as might be suspected.

THE GREEN ALGAE (CHLOROPHYCEAE)

ORDER SIPHONALES

18a Tubular free-living algae, somewhat branched. Sexual reproduction by non-motile eggs which are retained after fertilization within the oogonium. Both aquatic and terrestrial species.

Family VAUCHERIACEAE

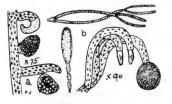

Figure 15

Fig. 15. a, *Vaucheria sessilis* D. C. with reproductive organs; b, *Dichotomosiphon tuberosum* Ernst.

The first species is often to be found thickly covering flowerpots, greenhouse beds or smooth damp ground in garden or field. It has been termed "green-felt".

18b A branching tubular thallus, parasitic within the leaves of plants of the family Araceae or the tissues of some mollusks.

Family PHYLLOSIPHONACEAE

Figure 16

Fig. 16. *Phyllosiphon arisari* Kuhn. a, filament; b, leaflet of Jack-in-the-pulpit with diseased spot. We do not often think of the algae as being parasitic but this is a good example.

19a Thallus without septa, the erect pinnately branched part attached by a prostrate rhizome. Gametes for reproduction are produced in the pinnate branches. Along both our Atlantic and Pacific coasts.

Family BRYOPSIDACEAE

Figure 17

Fig. 17. *Bryopsis corticulans* Setc.

Reproduction seems to be much more active in the spring. This marine coenocyte is too small to be located easily among its larger neighbors.

19b Much-branched tubular thallus with gametes born in specialized organs.

Family CODIACEAE

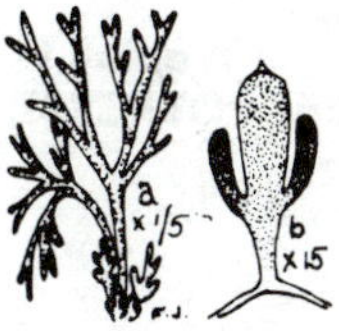

Figure 18

Fig. 18. *Codium fragil* Hariot. a, portion of plant; b, two gametangia.

All species are marine and many are limited to the warmer seas. This is another rather diminutive coenocyte.

——————————*——————————

20a Vegetative cells motile by means of flagella (some have non-motile resting periods). Figs. 19 to 24. ....21.

20b Vegetative cells not self-moving. Reproductive cells often motile. Figs. 25 to 52. .... **25**

ORDER VOLVOCALES

21a (a, b, c) Motile cells always solitary. Figs. 19 to 22. .... **22**

21b Motile cells in colonies. Figs. 23 and 24. .... **24**

21c Motile cells in one genus solitary and in the other colonial but distinguished from 21a and 21b by having many contractile vacuoles near the surface and processes of protoplasm extending to the wall. **Family SPHAERELLACEAE**

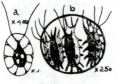

Figure 19

Fig. 19. a, *Sphaerella lacustris* Wittr.

Common in stone or earthenware cavities periodically filled with rain water.

b, *Stephanosphaera pluvialis* Cohn.

Ornamental urns in cemeteries are recommended as a likely place to look for *Sphaerella*. If that seems gloomy try a park; its the urn rather than the cemetery that counts.

22a Motile cells with no enclosing wall. With 2-4 or rarely 8 flagella and one eyespot. **Family POLYBLEPHARIDACEAE**

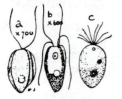

Figure 20

Fig. 20. a, *Stephanoptera gracilis* Smith; b, *Pyramimonas tetrarhynchus* Schm.; c, *Polyblepharides fragariiformis* Hazen.

An eyespot is always present in active individuals. The members of one genus, *Polytomella* are without chlorophyll.

22b Motile cells with an enclosing wall. Figs. 21 and 22. .... **23**

23a Wall consisting of two overlapping halves which push apart in reproduction. **Family PHACOTACEAE**

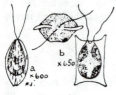

Figure 21

Fig. 21. a, *Phacotus lenticularis* Stein., b, *Pteromonas aculeata* Lemm.

These plants are surrounded by overlapping cells, somewhat similar to the diatoms. They always have just two flagella.

14

23b Wall continuous and not divided. With two or four flagella but
never more.   Family CHLAMYDOMONADACEAE

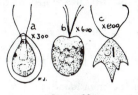

Figure 22

Fig. 22. a, *Chlamydomonas gloeocystiformis*
Dill.; b, *Platymonas elliptica* Smith; c, *Brachiomonas submarina* Bohlin.

A large family with several common species. Two is the most frequent number of
flagella, and four is the maximus number.

24a Colonies forming either a hollow sphere or. a flattened plate.
The individual cells and usually the colony embedded in gelatin.
                    Family VOLVOCACEAE

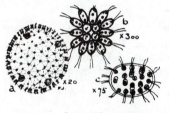

Figure 23

Fig. 23. a, *Volvox aureus* Ehr.; b, *Gonium pectorale* Muell.; c, *Pleodorina illinoiensis* Kof.

Volvox is well-known in theory as it
is often described in books on biology.
It is sometimes very abundant in permanent bodies of quiet water.

24b Colonies solid, made up of layers of four cells each. No gelatinous
covering.   Family SPONDYLOMORACEAE

Figure 24

Fig. 24. *Spondylomorum* sp.

Such minute algae as we are now considering
may appear in great abundance and shortly seem
to be wholly gone only to reappear when conditions
are again favorable.

25a Usually spherical or ovoid cells held together in colonies by a
gelatinous secretion in which the cells are embedded. These gelatinous colonies take various shapes and sizes. Figs. 25 to 28. . .26

25b Plants not forming gelatinous colonies. (A few members of the
Chlorococcales are an exception). Figs. 29 to 52. 29

ORDER TETRASPORALES

26a Cells attached by a stem which may be slender or broad and short. Solitary or in dendroid colonies.

Family CHLORANGIACEAE

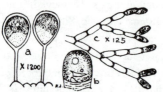

Figure 25

Fig. 25. a, *Stylosphaeridium stipitatum* G. & G.; b, *Malleochloris sessilis* Pash.; c, *Prasinochladus lubricus* Kuck.

The last species mentioned is marine. All the species are attached forms.

26b Cells not attached as in 26a. Figs. 26 to 28. 27

27a Vegetative cells with two or more long pseudocilia at their anterior end. Usually in microscopic or larger colonies.

Family TETRASPORACEAE

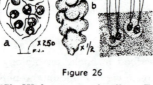

Figure 26

Fig. 26. a, colony of *Apiocystis brauniana* Nag.; b, colony and small part of *Tetraspora* sp.

These are found in quiet waters. Some are ephiphytic.

27b Without pseudocillia. Figs. 27 and 28. 28

28a Gelatinous sheath surrounding colony the same throughout; cells usually elongate.　　　　Family PALMELLACEAE

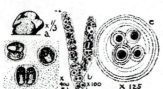

Figure 27

Fig. 27. a, *Palmella miniata* Lieb.; b, *Palmodictyon viride* Kutz.; c, *Gleocystis gigas* Lag.

A few members of this family grow out of water but most of them are aquatic.

28b Gelatinous sheath of each cell usually distinct; cells mostly spherical.　　　　Family COCCOMYXACEAE

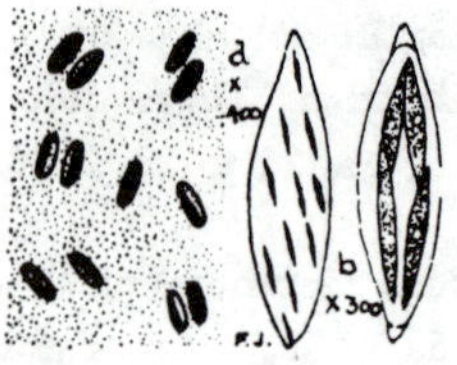

Figure 28

Fig. 28. a, *Coccomyxa dispar* Schm.; b, *Elkatothrix* spp.

Found in abundance among the floating vegetation in lakes, etc. Cell division is the only reproduction known.

29a (a, b, c) Single celled plants or multicellular colonies, but never forming filaments. (The water net (Fig. 35) is not a true filament). The cells take many unusual shapes. (A few colonies are embedded in a gelatinous mass; and another few expanding into a thallus). Figs. 30 to 37. 31

29b Plants highly multicellular, leaf-like or developing into hollow tubes or solid cylinders. Figs. 29, 38 and 39. 30

29c Not as in 29a or 29b but developing into a branched or unbranched filament. Figs. 40 to 52. 38

30a Chloroplasts star-shaped; reproduction only by non-motile asexual spores.   Order SCHIZOGONIALES
                                                    Family SCHIZOGONIACEAE

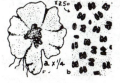

Fig. 29. *Prasiola mexicana* Ag. a, thallus; b, magnified surface view.

Some marine species belong here but the majority are inland forms. Some of the species are found on damp ground.

Figure 29

30b Chloroplasts not star-shaped; reproduction by zoospores and by motile isogametes. Figs. 38 and 39. 37

ORDER CHLOROCOCCALES

31a Cells solitary. Figs. 30 to 33. 32

31b Colonies of a definite number of coenobic cells. Figs. 34 to 37. 35

32a Elongated cells attached by a short stalk. Essentially solitary, though many plants often grow crowded together in a radiating cluster. Usually multinucleate. Reproduce by zoospores and by gametes, both biciliated.   Family CHARACIACEAE

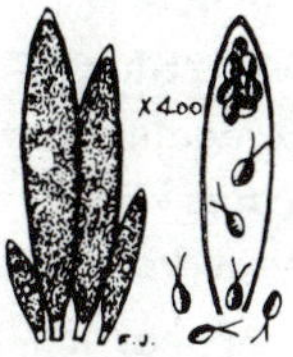

Fig. 30. *Characium angustatum* Br.

Grows attached to other algae, or upon stones or submerged wood. Vaucheria frequently bears plants of this group.

Figure 30

32b Not attached as in 32a. Figs. 31 to 33. 33

33a Cells multinucleate, spherical to elongate, often bearing an elongated root-like process. Reproduction by biciliated zoospores or gametes. Family PROTOSIPHONACEAE

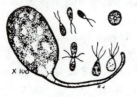

Fig. 31. *Protosiphon botryoides* Klebs.

Common on moist soil; often mixed with *Botrydium* with which it may be easily confused. *Protosiphon* contains starch; *Botrydium* never does; a starch test will settle any uncertainty.

Figure 31

33b Cells with but one nucleus. Figs. 32 and 33. . . . .**34**

34a Cells small, usually somewhat spherical, often variously ornamented but always rather symmetrical in shape. Aerial or aquatic, nucleus haploid. Family CHLOROCOCCACEAE

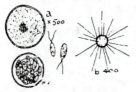

Fig. 32. a, *Chlorococcum humicola* Rab.; b, *Golenkinia radiata* Chod.

Often very abundant on damp soil and brickwork and in soil. A few species are aquatic.

Figure 32

34b Cells larger and unsymmetrical in shape. Usually found on or within other algae. Some species parasitic. Nucleus diploid.

Family ENDOSPHAERACEAE

Fig. 33. *Kentrosphaera bristolae* Smith.

This is a soil-living species. Many members of this family live parasitically on the mosses and flowering plants.

Figure 33

35a Solitary or colonial; when colonial, irregular with no fixed number of cells. Reproduction only by autospores (formation within the parent cell of several walled spores having the shape of the parent cell and which on release grow to its size). (A large family including many species of widely diversified shapes.)

Family OOCYSTACEAE

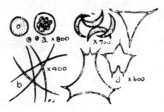

Fig. 34. a, *Chlorella variegatus* Bey.; b, *Ankistrodesmus falcatus* Ral.; c, *Selanastrum* sp.; d, *Tetraedron* spp.

Chlorella is interesting in that it frequently inhabits the tissues of animals in a cooperative way or as a parasite. The family is a very large one.

Figure 34

18

35b Colonies of definite arrangement. Figs. 35 to 37. 36

36a (a, b, c) Colony with an unchanging number of cells throughout its life; formed by the swarming of zoospores within the mother cell or other vesicle.   Family HYDRODICTYACEAE

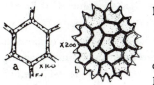

Fig. 35. a, *Hydrodictyon reticulatum* Lag. Water net; b, *Pediastrum boryanum* Men.

The net of *Hydrodictyon* takes an elongated sack-form, sometimes reaching a length of a foot or more.

Figure 35

36b Unit, a colony formed by the union of auto spores after their liberation. Number of cells from 4 to 128.

Family COELASTRACEAE

Fig. 36. *Coelastrum microporum* Nag.

It is an inhabitant of fresh-water lakes, and ditches but is seldom very abundant.

Figure 36

36c 2-4 or 8-cells (occasionally more) each with single nucleus arranged parallel, radially or in other definite form.

Family SCENEDESMACEAE

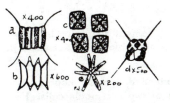

Fig. 37. a, *Scenedesmus quadricauada* Breb.; b, *Scenedesmus obliquus* Kutz.; c, *Crucigenia quadrata* Mor.; d, *Tetrastrum elegans* Pla.; c, *Actingstrum hantzschii* Lag.

This rather large family takes many forms. The algae scoured from the walls of fish bowls often belong here.

Figure 37

ORDER ULVALES

37a Mature thallus a solid cylinder several cells thick.

Family SCHIZOMERIDACEAE

Fig. 38. *Schizomeris leibleinii* Kutz.; a, cross section; b, longitudinal view of part of thallus.

Usually in clear water; not abundant but widely distributed.

Figure 38

19

37b Thallus a hollow tube, or ribbon-like, or flattened leaf like. In each case the wall only one or two cells thick.

Family ULVACEAE

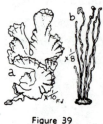

Fig. 39. a, *Ulva lactuca* L. Sea Lettuce; b, *Entero-morpha intestinalis* Grev.

Sea lettuce attains a height of nearly a foot. Its vivid brilliant green makes a sharp and pleasing contrast to the brown and red algae with which it grows.

Figure 39

38a (a, b, c) Zoospores and male reproductive cells with a ring of many flagella at apical end. With apical caps indicating where cell division has occurred. Filaments simple or branched.

Order OEDOGONIALES
Family OEDOGONIACEAE

Fig. 40. a, *Oedogonium* spp.; b, *Bulbo-chaete mirabilis* Wit.

These are naturally attached plants though broken parts or even large masses are found freely floating. A basal cell develops a holdfast for attachment.

Figure 40

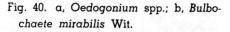

38b Zoospores with two or four flagella. Filament simple or branched (only a singular spherical cell or irregular colony of such cells in *Protococcus*). Figs. 41 to 49. 39

38c No zoospores. No flagellated reproductive spores. Reproduction by a zygospore formed by the union of two non-motile cells. Filament always unbranched. Figs. 50 to 52. 46

39a Cells with but one nucleus and one chloroplast. Figs. 41 to 47...40

39b Cells with more than one nucleus and with several discoid chloroplasts. Figs. 48 and 49. 45

ORDER ULOTRICHALES

40a (a, b, c) Cells solitary, in irregular groups or in plate-like colonies. Figs. 44 and 45. 43

40b Cells in colonies of branched filaments. Figs. 46 to 49. 44

40c Cells in colonies of unbranched filaments. Figs. 41 to 43. 41

41a Filament wall composed of a series of overlapping H-shaped pieces. Chloroplast lobed but often indefinite as to shape.

Family MICROSPORACEAE

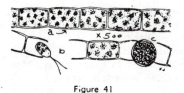

Figure 41

Fig. 41. *Microspora willeana* Wittr.; a, filament; b, zoospore and H-pieces; c, aplanospore being released from filament.

These plants are at their best in the spring where they are common in small quiet waters.

41b Without H-pieces in wall of filament. Figs. 42 and 43. 42

42a Each cell of filament enclosed in its own cellulose wall making the filament wall appear stratified.

Family CYLINDROCAPSACEAE

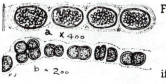

Figure 42

Fig. 42. *Cylindrocapsa geminella* Wolle.; a, young filament; b, older filament.

Try pools and ditches in late spring if you wish to find it.

42b Filament wall not stratified, the single chloroplast plate-like and lying near the cell wall. **Family ULOTRICACEAE**

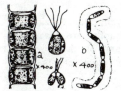

Figure 43

Fig. 43. a, *Ulothrix zonata* Kutz.; b, *Geminella spiralis* Smith.

These plants are normally attached and may be very abundant in slow-running water.

43a Cells solitary or sometimes in irregular masses of 2 to 50 cells. No zoospores or gametes known. **Family PROTOCOCCACEAE**

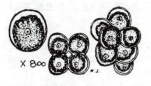

Figure 44

Fig. 44. *Protococcus viridis* Ag.

Very common on tree trunks, etc. This and a few other species are distributed world-wide, and it is one of our ever-present forms.

43b Part or all of the cells bear thread-like setae. Cells solitary or in plate-like or branching colonies. Both zoospores and gametes formed. **Family COLEOCHAETACEAE**

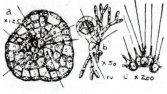

Figure 45

Fig. 45. a, *Coleochaete scutata* Breb.; b, *Coleochaete pulvinata* Br.; c, *Chaetosphaeridium* sp.

Commonly found attached to higher aquatic plants or to other algae. Most of the species reproduce sexually.

44a Zoospores and gametes produced in any vegetative cell and never in specialized cells. The ends of branches are often prolonged into colorless setae. **Family CHAETOPHORACEAE**

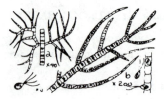

Figure 46

Fig. 46. a, *Draparnaldia plumosa* Agar.; b, *Stigeoclonium lubricum* Katz.

Clear cool running water is its favorite habitat. The abundant branching is characteristic.

44b Zoospores and gametes produced in specialized cells, differing from vegetative cells. Chloroplasts often several in one cell. **Family TRENTEPOHLIACEAE**

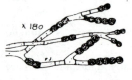

Figure 47

Fig. 47. *Ctenocladus circinnatus* Borzi.

The species of this family are less abundant than other members of the order. It lives as an epiphyte on other plants.

ORDER CLADOPHORALES

45a Cells very long, 12 to 50 times their width. Chloroplasts numerous in transverse bands. Gravel pits and flooded areas are said to be good places to look for it.

Family SPHAEROPLEACEAE

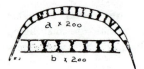

Figure 48

Fig. 48. *Sphaeroplea annulina* Ag.; a, single celled young plant; b, part of one cell of filament.

It is found in pools, often in connection with gravel.

45b Cells shorter, seldom with length more than eight times the width. Chloroplasts not in transverse band. Family CLADOPHORACEAE

Fig. 49. a, *Cladophora glomerata* Kutz. This family has both fresh water and marine species and is world-wide in its distribution. There are many species some of which form ball-shaped colonies.

Figure 49

ORDER ZYGNEMATALES

46a Cells cylindrical, united in filaments. Reproduce by conjugation process in which protoplasts have no chance to escape to exterior. Family ZYGNEMATACEAE

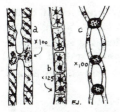

Fig. 50. a, *Spirogyra* sp.; b, *Zygnema* sp.; c, *Mougeotia* sp.

No one can get very far in nature lore without hearing about Spirogyra. There are many species, some with but one spiral chloroplast to the cell, others with several intertwined.

Figure 50

46b Cells of various shapes, usually solitary or occasionally united into simple filaments. Figs. 51 and 52. 47

47a Cell walls with vertical pores. In conjugation the protoplast escapes from the surrounding walls. Family DESMIDIACEAE

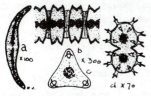

Fig. 51. a, *Closterium* sp.; *Desmidium* sp.; b, side view, c, end view; d, *Schizocanthum* sp.

These highly interesting plants somewhat resemble diatoms but do not have stiff walls.

Figure 51

47b Cell walls without pores. The protoplast confined by walls in conjugation. Family MESOTAENIACEAE

Fig. 52. a, *Cylindrocystis diplospora* Lund. Grows on damp soil; b, *Netrium digitus* I. & R|

These are free-floating fresh-water species which reproduce by cell division and less frequently conjugation.

Figure 52

————————*————————

48a (a, b, c) Plants yellow, yellow-green or golden-brown. Mostly microscopic species. Figs. 53 to 83. 49

48b Plants brown or greenish-brown. Usually large plants, some very large. Most species marine. Male gametes pear-shaped with two flagella on the side. Figs. 84 to 100. (The Brown Algae). 75

48c Plants red or purplish-green. Usually medium to large sized plants. Most species marine. No motile reproductive cells. Figs. 101 to 116. (The Red Algae). 86

49a Single celled plants (sometimes united in chains or circular figures) with outer wall glass-like (of silica) and always composed of two overlapping parts (frustules) like a box and its cover. Contents yellow or yellowish-brown. (The Diatoms). Figs. 53 to 67. 50

49b Not as in 49a. Figs. 68 to 83. 61

THE DIATOMS (BACILLARIEAE)

50a Frustules (outer walls) circular, polygonal or irregular in outline. Markings radiating from a center point. Figs. 53 to 56. 51

50b Frustules bilaterally symmetrical or irregular in surface view. Markings always arranged around a line and never around a point. Figs. 57 to 67. 54

ORDER CENTRALES

51a Frustules without horns or prominent spines; shape a short cylinder. Most species disk-like though some are taller than broad.
Family COSCINODISCACEAE

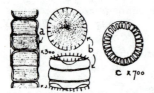

Figure 53

Fig. 53. a, *Melosira varians* Ag. grows in long filaments; b, *Stephanodiscus niagarae* Ehr. valve view; girdle view; c, *Cyclotella meneghiniana* Kutz.

"Frustule" seems to be about the best term for the entire wall or "shell" of a diatom. The word is sometimes used to include the contents, also.

51b Frustules with horns or prominent spines (some long cylindrical species lack spines or horns). Figs. 54 to 56. 52

52a Frustules elongated cylinders with many intercalary bands between the girdles.
Family RHIZOSOLENIACEAE

Figure 54

Fig. 54. *Rhizosolenia eriensis* Smith.

"Valve" is used to refer to the face view or side of the frustule. The overlapping half on the wall is the hypotheca; while the smaller half is the epitheca.

52b Valves (face of frustule) with two or more thick horns or projections, the two families here-in treated are bilaterally symmetrical or asymmetrical but have the ornamentation radially arranged. Figs. 55 and 56. 53

53a Valves with spines or elevations at the angles.
Family BIDDULPHIACEAE

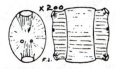

Fig. 55. *Biddulphia laevis* Ehr.

A diatom so placed as to show the edges of the overlapping halves is said to present a girdle view.

Figure 55

53b Valves with internal partitions vertical to the valve face.
Family ANAULACEAE

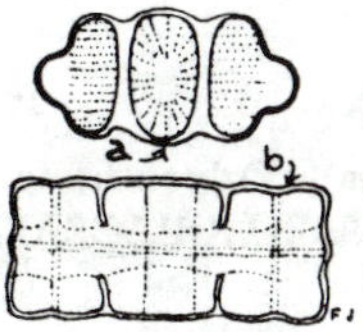

Fig. 56. *Terpsinoe americana* Ralfs. a, valve view; b, girdle view.

The very fine markings on the valves were once much used to test microscopes.

Figure 56

ORDER PENNALES

54a The two valves always unlike, one having a raphe and the other a pseudoraphe. There are no internal septa except in one genus which has incomplete longitudinal ones. Family ACHNANTHACEAE

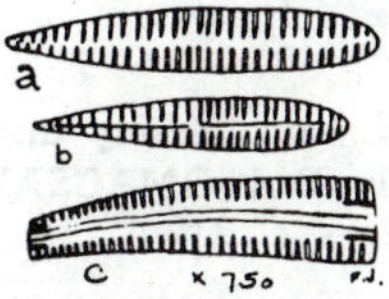

Fig. 57. *Rhoccosphenia curvata* Grun.; a, other valve or epitheca; b, inner valve of hypotheca; c, girdle view.

Diatoms are world wide in their distribution.

Figure 57

54b Not as in 54a. Figs. 58 to 67. 55

55a (a, b, c) The frustules usually elongate; the two valves alike in that each has a raphe or each member of the pair has a pseudoraphe. Figs. 58 to 62. 56

55b Both valves with a true raphe which lies toward the middle of the valve not in the marginal keel. Figs. 63 to 65. 59

55c Both valves alike; each raphe somewhat concealed in a keel at one or both sides of the valve. Figs. 66 and 67. 60

56a Valves bow-shaped, the raphe or pseudoraphe lying toward the concave side. Girdle view rectangular or wedge-shaped.
Family EUNOTIACEAE

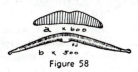

Fig. 58. a, *Eunotia pectinalis* Rab.; b, *Ceratoneis arcus* Kutz.

A bit of slime from the bottom of any watercourse is practically certain to be filled with diatoms.

Figure 58

56b Not as in 56a. Figs. 59 to 62. 57

57a Valves wedge-shaped with transverse septa transversely asymmetrical.
Family MERIDIONACEAE

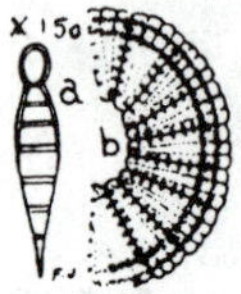

Fig. 59. *Meridion constrictum* Ral.; a, valve view; b, girdle view.
Many species of diatoms are known only as fossil forms.

Figure 59

57b Not wedge-shaped. Figs. 60 to 62. 58

58a (a, b, c) With longitudinal internal septa (partitions). Valves transversely symmetrical; usually long.
Family TABELLARIACEAE

Fig. 60. *Tabellaria fenestra* Kutz.; a, typical arrangement of colony; b, valve view; c, girdle view.
Immense deposits of the frustules of diatoms are found and have wide uses comcercially. This is known as diatomaceous earth.

Figure 60

58b With transverse internal septa; symmetrical both transversely and longitudinally.
Family DIATOMACEAE

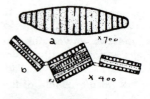

Fig. 61. *Diatoma vulgari* Bory.; a, valve view; b, girdle view and typical colonial form.
Diatomaceous earth is used for polishing, as a filter in refining sugar, for heat insulation, to give body to dynamite, in brick and cement work, and in many other ways.

Figure 61

58c With no internal septa. Both transversely and longitudinally symmetrical.
Family FRAGILARIACEAE

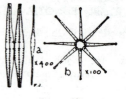

Fig. 62. a, *Fragilaria* sp.; b, *Asterionella gracillima* Heib.
The free movement of diatoms in the water attracts attention. Of course, it must be remembered that the apparent speed is magnified the same as the size.

Figure 62

59a (a, b, c) Valves symmetrical both transversely and longitulinally. Valves of many shapes. Family NAVICULACEAE

Figure 63

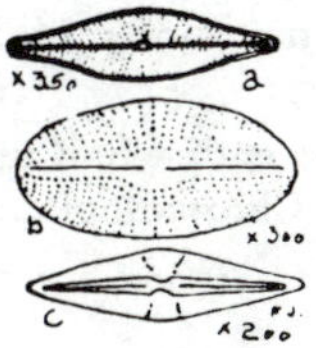

Fig. 63. a, *Navicula rhyncocephala* Kutz.; b, *Diploneis elliptica* Cleve.; c, *Frustulia rhomboides* De T. Streaming protoplasm seems to offer the best solution as to how a diatom propels itself.

59b Valves symmetrical longitudinally but not symmetrical transversely. Family GOMPHONEMATACEAE

Figure 64

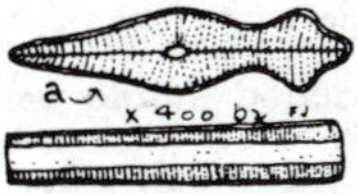

Fig. 64. *Gomphonema acuminatum* Ehr.; a, valve view; b, girdle view, a rather common fresh-water form.

Reproduction of diatoms is most commonly by cell division. Each half takes one valve and grows a new wall for the other side.

59c Valves symmetrical transversely but not so longitudinally. Family CYMBELLACEAE

Figure 65

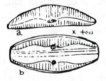

Fig. 65. *Amphora ovalis* Kutz.; a, valve view; b, girdle view.

This species is unusually large. The family is a large one, too. Since the newly formed wall part fits inside the old half some of the individuals become smaller each generation.

60a With a single eccentric keel lying near one lateral margin. Family NITZSCHIACEAE

Figure 66

Fig. 66. *Hantzshia* sp.

Two diatoms sometimes shed their outer shells and fuse to later separate and grow new shells. This is conjugation. Is is by this means that they regain their normal size.

60b With two keels, one near each margin of the valve. Family SURIRELLACEAE

Figure 67

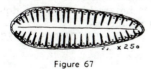

Fig. 67. *Surirella splendida* Kutz.

Cell division of diatoms most usually takes place around midnight.

---*---

61a Motile cells with flagella inserted in a transverse furrow; one flagellum wraps around the cell transversely; the other extends vertically backward. Contents or organism golden-brown or colorless. Figs. 68 to 71. 63

61b Plants without a transverse groove. Figs. 72 to 83. 62

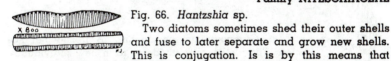

62a Motile cells with two unequal flagella at anterior end. Discoid chromatophores yellowish-green; without pyrenoides; stored food, oil, not starch. The cell walls of many species are made of H-shaped pieces which overlap. Figs. 72 to 79. 66

62b Motile cells with one or two flagella at the anterior end; when two they may be either equal or unequal in length. Chromatophores are a distinctive golden-brown. Figs. 80 to 83. 71

CLASS DINOPHYCEAE

This group of animal-like plants or plant-like animals (zoologists and botanists both claim them) are numerous—some 1000 species have been described but many points concerning their classification seems to be still questioned. They are likely of much less importance than other plants that have been omitted but some proposed orders are given.

63a Motile, free swimming in vegetative state. Figs. 69 to 71. 64

63b Vegetative stage non-motile*, free-floating or attached single cell-ed. Reproduce only by zoospores or autospores.

Order DINOCOCCALES

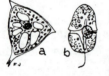

Fig. 68. *Tetradinium minus* Pash. a, mature vegetative plant; b, zoospore.

We have divided this class only to orders since there is still a great deal to learn about it before its classification is certain.

Figure 68

64a Cell enclosed in a hard wall composed of a definite number of plates. Figs. 70 and 71. 65

64b Cell naked or if within a wall it is not divided into plates. Mostly solitary; a few colonial; mostly marine.   Order GYMNODILIALES

X 200

Fig. 69. a, *Gymnodinium fascum* Ehr. A fresh-water species. The motile cells divide to form new individuals.

Figure 69

65a Wall divided vertically into two opposite halves (each with a definite number of plates). Known species all marine.

Order DINOPHYSIDALES

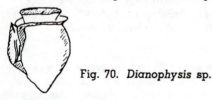

Fig. 70. *Dianophysis* sp.

Figure 70

*The non-motile Dinophyceae have been divided into four orders but since three of these have but one or two known species, they are not included in the key.

65b Wall not separated vertically into two halves but divided into a definite number of plates. Wholly marine.

Order PERIDINIALES

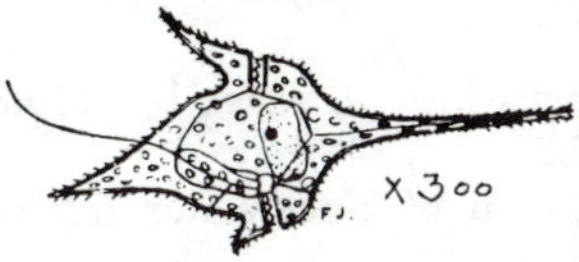

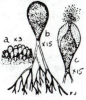

Fig. 71. *Ceratium hirundinella* Sch.

There are a few fresh water species but most of them are marine.

Figure 71

CLASS HETEROKONTAE

66a (a, b, c) Plant single celled with bulb-like multinucleate head and branching root-like parts, growing in soil.

Order HETEROSIPHONALES
Family BOTRYDIACEAE

Fig. 72. *Botrydium granulatum* Grev.; a, group of plants; b, single plant; c, zoospores being discharged from above-ground part.

These very interesting little plants are often common on the smooth ground of gardens and paths.

Figure 72

———————————*———————————

66b Cells arranged in filaments, either simple or branched. Figs. 77 to 83. . . . .**70**

66c Not like either 66a or 66b. Usually single celled. Surrounded by a wall. Figs. 73 to 76. . . . .**67**

ORDER HETEROCOCCALES

67a Cells always solitary. Figs. 75 and 76. . . . .**69**

67b Cells colonial or solitary. Figs. 73 and 74. . . . .**68**

68a Cells cylindrical often in dendroid colonies; free-floating or epiphytic.

Family OPHIOCYTIACEAE

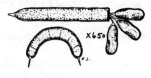

Fig. 73. a, *Ophiocytium* spp.

Found in plankton and in pools. Spores form within the cells and escape to develop into new individuals.

Figure 73

68b Various shapes other than cylindrical; but one nucleus; cells small with one or a few chromatophores; reproduce only by autospores.
Family BOTRYOCOCCACEAE

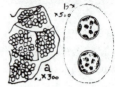

Fig. 74. a, *Botryococcus braunii* Kutz.; b, Chlorobotrys reglaris Boh.
Found in the plankton of clear-water lakes.

Figure 74

69a Cells usually with but one nucleus; large and with many chromatophores. Reproduce by both zoospores and autospores.
Family HALOSPHAERACEAE

Fig. 75. a, *Leuvenia natans* Gard. Grows in the film covering water.; b, *Botrydiopsis arhiza* Borzi. Grows on the surface of the ground.

Figure 75

69b Attached epiphytic forms often multinucleate with one to several chromatophores. Family CHLOROTHECIACEAE

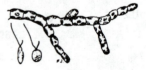

Fig. 76. a, *Characiopsis pyriformis* Bor.; b, *Peroniella planctonica* Smith.
Many of the species of algae have but little apparent relation to man and his schemes.

Figure 76

ORDER HETEROTRICHALES

70a Filaments branched and multicellular. Family MONOCILIACEAE

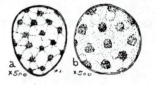

Fig. 77. *Monocilia viridis* Gern.
The plants of this family are soil-growing species, some being found even at a depth of a few feet.

Figure 77

70b Filaments unbranched. No starch grains.
Family TRIBONEMATACEAE

Fig. 78. *Tribonema* spp.; a, living plant; b, cell wall.
This family is very common in stagnant water. In early spring it is particularly abundant.

Figure 78

CLASS CHRYSOPHYCEAE

71a Cells motile during entire vegetative state or if amoeboid such condition is temporary. Figs. 80 to 83. 72

71b Vegetative cells always non-motile and united in gelatinous colonies.

ORDER CHRYSOCAPSALES

Very much branched colonies in which cell division occurs only near the apices. **Family HYDRURACEAE**

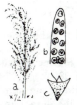

Fig. 79. *Hydrura foetidus* Kir.; a, branch of thallus; b, tip of thallus; c, zoospore.

Common on rocky bottoms of cold mountain streams. As the species name suggests, it reveals itself by its bad odor.

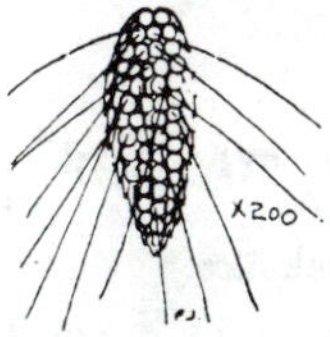

Figure 79

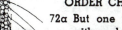

ORDER CHRYSOMONADALES

72a But one apical flagellum present; ornamented with scales and long spines.

Family MALLOMONADACEAE

Fig. 80. *Mallomonas* sp.

These plants are found in the floating growth of our fresh-water lakes.

Figure 80

72b With 2 apical flagella. Figs. 81 to 83. 73

73a Flagella of equal length. Figs. 82 and 83. 74

73b Flagella unequal in length. Outer surface the same throughout. **Family OCHROMONADACEAE**

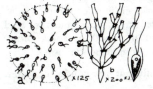

Fig. 81. a, *Urognelopsis americana* Lemm. Very common in reservoirs and lakes. Colonies may contain a thousand cells.; b, *Dinobryon sertularia* Ehr.; colony and individuals.

This species and other members of the genus are common in the plankton of fresh-water lakes.

Figure 81

74a Outer surface of the cells interrupted with scales of silica.

Family SYNURACEAE

Fig. 82. *Synura uvella* Ehr.

Inhabits the plankton of lakes and smaller bodies of water. The flagella are equal in length. The individuals possess no eyespot.

Figure 82

Figure 83

74b Outer surface of cells the same throughout.

Family SYNCRYPTACEAE

Fig. 83. *Syncrypta volvox* Ehr.

Many of the motile algae have a red eyespot but this species is not thus equipped.

THE BROWN ALGAE (PHAEOPHYTA)

75a (a, b, c) With alternation of two similar generations. Submerged marine plants. Figs. 84 to 88. 76
75b With alternation of two quite different generations. Marine plants partially exposed, especially at low tide. Figs. 89 to 98. 79
75c Without alternation of generations (only a diploid generation known). Marine plants normally growing between the tide lines and exposed during low tide. Figs. 99 and 100. 85

CLASS ISOGENERATAE

76a Thallus usually or for greater part filamentous or cylindrical. Figs. 84 to 86. 77
76b Thallus usually or for greater part flattened or disk-like. Figs. 87 and 88. 78

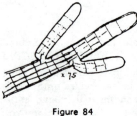

Figure 84

77a (a, b, c) Greater part of thallus with tiers of elongated vertical cells. Thallus with one large apical cell at tip from which all growth starts.

Family SPHACELARIACEAE

Fig. 84. *Sphacelaria californica* Sauv.

Members of this family are marine and constitute much of the food of some plant-eating fish and other marine animals.

77b Thallus filamentous and branched; usually but one cell in diameter. Reproduce by zoospores and isogametes.

Family ECTOCARPACEAE

Figure 85

Fig. 85. *Ectocarpus cylindricus* Saund.; a, portion of plant; b, gametangium; c, a sporangium.

These plants are usually attached to coarser algae. They are marine.

77c Upper part of thallus usually one cell in diameter; lower part with elongated cells in transverse tiers.

Family TILOPTERIDACEAE

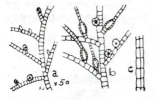

Fig. 86. *Haplospora globosa* Kjel.; a, upper part of sporophyte; b, upper part of gametophyte; c, lower part of thallus.

Figure 86

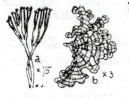

78a Alternating generations differ widely in size and form; asexual plants often disk-like.

Family CUTLERIACEAE

Fig. 87. *Cutleria multifida*; a, sexual plant; b, young asexual plant.

It is found in the warmer parts of the north Atlantic.

Figure 87

78b Alternating generations quite similar in general appearance.

Family DICTYOTACEAE

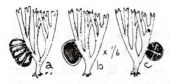

Fig. 88. *Dictyota dichotoma* Lam.; a, female plant; b, male plant; c, asexual plant. (The reproductive organs are enlarged out of proportion to the thali in the drawing.)

One investigator reports a fair sized plant of this species which produced over 500 million sperms at one time. Many of nature's creatures are thus highly prolific.

Figure 88

CLASS HETEROGENERATAE

79a Thallus built of thread-like branching filaments often attached to each other at their base. Figs. 89 to 92. 80

79b A thallus with true parenchymatous tissue is formed by intercalary longitudinal division. Figs. 93 to 98. 82

80a A larger sporophyte of sporangia bearing branching filaments alternates with a microscopic monoecious gametophyte which reproduces by isogametes. Figs. 91 and 92. .81

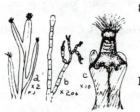

Figure 89

80b Sporophyte with each tip terminating in a tuft of hairs. Microscopic monoecious gametophyte reproduces by sperms and eggs. Order SPOROCHNALES
Family SPOROCHNACEAE

Fig. 89. *Carpomitra cabrerae* Kutz.; a, part of sporophyte; b, gametophyte; c, enlarged tip of sporophyte.

The brown algae are exceedingly abundant and highly important plants.

80c Each sporophyte ending in a single filament. The microscopic gametophyte produces sperms and eggs. Order DESMARESTIALES
Family DESMARESTIACEAE

Figure 90

Fig. 90. *Desmarestia herbacea* Lamx.; a, portion of thallus; b, female gametophyte; c, male gametophyte.

The sporophyte attains a length of several feet. Rather common along the Pacific coast.

ORDER CHORDARIALES

81a Sporophyte erect cylindrical single or branched. Gametophyte filamentous; microscopic. Family CHORDARIACEAE

Figure 91

Fig. 91. *Mesogloia vermiculata* Le J.; a, sporophyte; b, gametophyte.

This plant belongs to the north Atlantic, where it grows among the larger members of the class.

81b Sporophyte spreading over surface of attachment. Globular, solid or hollow. Family LEATHESIACEAE

Figure 92

Fig. 92. *Leathesia difformis* Aresc.; a, sporophyte; b, sporangia; c, gametophyte.

A northern species found in both the Atlantic and Pacific oceans.

82a (a, b, c) Sporophyte of medium size, or unknown, parenchymatous tissue all much the same; grows by intercalary cell division; gametophyte macroscopic (may be seen with un-aided eye). Figs. 93 to 95. 83

82b Sporophyte with definite holdfast, stipe and blade; often very large. Growth due to intercalary meristem. Gametophyte microscopic and producing sperms and eggs. (The Kelps) In this order are found the largest algae known. Figs. 96 to 98. 84

82c Thallus cylindrical, much branched; new growth from a single apical cell. Gametophyte microscopic; produces isogametes.

Order DICTYOSIPHONALES
Family DICTYOSIPHONACEAE

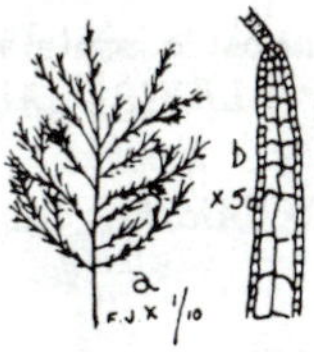

Fig. 93. *Dictyosiphon foeniculaceus* Kutz.; a, sporophyte; b, tip of branch.

Found on both the Atlantic and Pacific coasts.

Figure 93

ORDER PUNCTARIALES

83a Apparently only the gametophyte known; hairs in groups; gametangia pallisade-like not reaching beyond the surface.

Family SCYTOSIPHONACEAE

Fig. 94. *Scytosiphon lomentaria* Agar.; a, part of gametophyte; b, section through a gametangium.

Widely scattered on all our coasts.

Figure 94

83b Both sporophyte and gametophyte small but macroscopic when mature; hairs present. Sporangia in elevated sori.

Family ASPEROCOCCACEAE

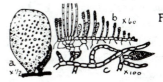

Fig. 95. *Soranthera ulvoidea* P. & R.; a, sporophyte covered with sori; b, section through sorus; c, gametophyte.

Common along the Pacific coast.

Figure 95

ORDER LAMINARIALES

84a (a, b, c) Sporophyte simple, with unbranched stipe; the lamina sometimes cut or divided. Holdfast often branched, sometimes disk-like. Family **LAMINARIACEAE**

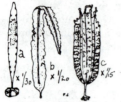

Fig. 96. a, *Laminaria saccharina* Lam.; b, *Cymathaere triplicata* Ag.; c, *Costaria costata* Saund.

The family is a large one and some of the species attain considerable size. It has importance as a food source in Japan.

Figure 96

84b Sporophyte simple or irregularly branched. Sori almost completely covers both sides of lamina. Family **ALARIACEAE**

Fig. 97. a, *Egregia menziesii* Ares.; b, *Alaria crassifolia* Kjel.

Widely distributed in both the oceans.

The holdfast by which some of these large species are attached is often a complicated structure.

Figure 97

84c Sporophyte more or less compound with stipe divided or branching and few to many lamina. Family **LESSONIACEAE**

Fig. 98. a, *Macrocystis pyrifera* Ag. This very common west coast species may attain a length of 175 feet; b, *Postelsia palmaeformis* Rupr.

This so called sea-palm though only 15 to 20 inches high attracts much interest by its tree-like appearance.

Figure 98

ORDER FUCALES

85a Thallus flattened with all branching in one plane. Female gametangium producing 8-4-2-1 large viable eggs. Family **FUCACEAE**

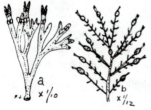

Fig. 99. a, *Fucus vesiculosus* L.; b, *Ascophyllum nodosum* Le J.

This is a standard laboratory plant. It has some other uses also such as being a source of iodine.

Figure 99

85b Thallus rounded and branching from all sides, female gametangium producing but one large viable egg. Family **SARGASSACEAE**

Fig. 100. a, *Sargassum* sp.; b, *Pycnophycus tuberculatus* Kutz.

Some 150 species of *Sargassum* have
been named. It is a familiar sight along
the beach of the Gulf of Mexico and other
warm waters where its plants are washed
up on the shores.

Figure 100

THE RED ALGAE (RHODOPHYTA)

86a Thallus enlarges by intercalary growth; zygote divides directly
into carpospores. A few species are unicellular.

Fig. 101. a, *Porphyra perforata* Ag. Common along
both our coasts; b, *Porphyridium cruentum* Nag. Appears as a red crust on damp earth.

Order BANGIALES
Family BANGIACEAE

Figure 101

86b Thallus grows only at tip; carpospores arise indirectly from the
zygote. Figs. 102 to 116. 87

87a Sporophyte generation represented only by the zygote. Male and
female gametophytes well developed. Figs. 104 to 107. 91

87b A definite sporophyte generation alternating with male and female gametophytes. Figs. 102, 103 and 108 to 116. 88

88a Without an auxiliary cell (separate vegetative cell which receives
the zygote nucleus in its migration from the carpogonium); the
sporophyte developing directly from the carpogonium.

Family GELIDIACEAE

Fig. 102. *Gelidium corneum* Lam.
It is the members of this family that furnish agar-agar,
the most valuable product coming from any of the algae.
The possibilities for food and other commercial products from marine algae are large.

Figure 102

88b Auxiliary cell present. Figs. 103 and 108 to 116. 89

89a Auxiliary cell arising on a special filament. Figs. 108 to 110. ...92

89b Not as in 89a. Figs. 103 and 111 to 116. 90

90a Auxiliary cell arising from a vegetative cell of the gametophyte. Figs. 111 to 113. 94

90b Auxiliary cell a special cell prepared before fertilization.

Order RHODYMENIALES

Carpogonial branch three celled.   Family RHODYMENIACEAE

Figure 103

Fig. 103. *Rhodymenia palmata* Grev.

This plant is an article of food in some regions and is sometimes used for chewing, like tobacco.

90c A special auxiliary cell prepared after fertilization. Figs. 114 to 116. 95

ORDER NEMALIONALES

91a (a, b, c, d) Thallus without central axis and without whorls of branches.

Family CHANTRANSIACEAE

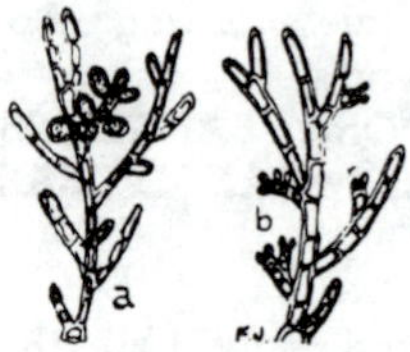

Fig. 104. *Acrochaetium rhipidandrum* Kyl.; a, female gametophyte; b, male gametophyte.

This family contains both freshwater and marine species.

Figure 104

91b Thallus with a main axis surrounded by thick whorled tufts of finer filaments.

Family BATRACHOSPERMACEAE

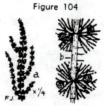

Fig. 105. *Batrachospermum moniliforme* Roth.; a, part of plant; b, whorls and reproductive organs. This is a fresh water species.

Figure 105

91c Thallus of branching filaments for most part only one cell in diameter. The zygote grows an uncovered tuft of surrounding filaments.   Family HELMINTHOCLADIACEAE

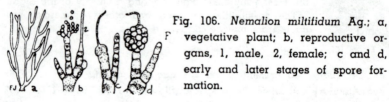

Fig. 106. *Nemalion miltifidum* Ag.; a, vegetative plant; b, reproductive organs, 1, male, 2, female; c and d, early and later stages of spore formation.

Figure 106

91d Thallus of slender colorless filaments dichotomously branched Carposporophyte surrounded by a pericarp.

Family CHAETANGIACEAE

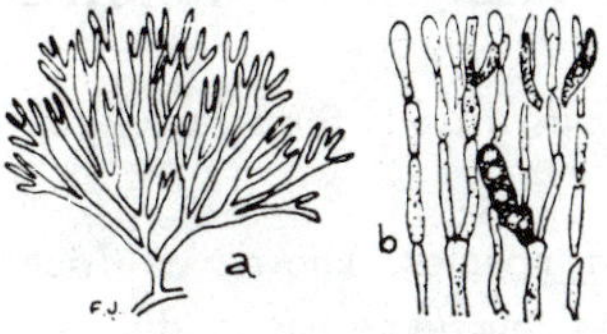

Figure 107

Fig. 107. *Scinaia furcellata* Biv.; a, part of thallus; b, cells within filament.

This species is found in the Atlantic and Mediterranean.

ORDER CRYPTONEMIALES

92a Plants much incrusted with lime; antheridia in conceptacles with a pore outlet.

Family CORALLINACEAE

Figure 108

Fig. 108. a, *Lithothamnion* sp.; b, *Corallina mediterranea*.

Members of this family are often so encrusted and stiff as to appear coral like, hence the name.

92b Not incrusted with lime. 93

93a Thallus grows only at terminal cells; no procarp. Carpogonium and auxiliary cells on special multicellular branches.

Family DUMONTIACEAE

Fig. 109. *Dudresnaya purpurifera* Ag.; part of plant.

Reproduction in the red algae is usually a complicated process. Some exceedingly delicate and beautiful plants belong to this order.

Figure 109

93b Carpogonial branches and the auxiliary cell branches gathered into bodies on the surface of the thallus.

Family SQUAMARIACEAE

Fig. 110. *Peyssonnelia squamaria* Dec.

These plants are sometimes encrusted with lime.

Figure 110

39

ORDER GIGARTINALES

94a (a, b, c) Basal cell becomes the auxiliary cell. Thallus fleshy, dichotomously branched. Family GIGARTINACEAE

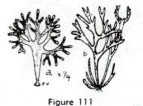

Fig. 111. a, *Chondrus crispus* Stac.; b, *Gigartina* sp.

The first named species, known as Irish Moss is much used for several food dishes.

Figure 111

94b A daughter cell of the basal cell becomes the auxiliary cell, appearing after fertilization. Family RHODOPHYLLIDACEAE

Fig. 112. *Cystoclonium purpurascens* Kutz.; a, part of frond; b, auxiliary cell and goniomoblast.

Figure 112

94c Auxiliary cell large, surrounded by several carpogonial cells. Many nutritive cells present. Family RISSOELLACEAE

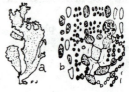

Fig. 113. *Rissoella verruculosa* Ag.; a, frond; b, cross section showing auxiliary cell.

The sperms of the red algae are non-motile and float at random in the water.

Figure 113

ORDER CERMIALES

95a Thallus leafy, sometimes with a midrib.

Family DELESSERIACEAE

Fig. 114. a, *Grinnellia americana* Har.; b, *Delesseria sanguinea.*

Many of the most beautiful of the red algae belong in this rather large family.

Figure 114

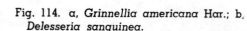

95b Thallus of branched filaments. Figs. 115 and 116. 96

96a Filaments monosiphonous (but one cell in diameter).
Family CERAMIACEAE

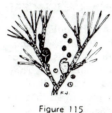

Figure 115

Fig. 115. *Callithamnion corymbosum* Lyn.; branch of tetrasporic generation.

Alternation of a sexual generation with an asexual one is the rule with red algae.

96b Filaments polysiphonous (more than one cell in diameter).
Family RHODOMELACEAE

Figure 116

Fig. 116. a, *Dasya elegans* Ag.; b, *Polysiphonia violaceae* Har.

These fine-cut plants are very attractive, especially when floating in water.

THE LICHENS (Phylum Lichenes)

The situation here is different from all other parts of the plant kingdom. Technically the Lichens are not in themselves individual plants but plant corporations or partnerships, the individual members of which already have places elsewhere in the families of fungi and algae. The keys that follow consider these plants, then, as corporate bodies and not the individual components. Some botanists prefer to thus give the Lichens a place of their own in the classification. Others wish to distribute them among the fungi and algae to which their component symbionts belong. Those of such mind may disregard these keys.

1a Lichens in which the spores are borne on basidia (clubs). They take the form of bracket fungi.　　　Order BASIDIOLICHENES
This order including but a limited number of tropical species, no further division of them will be attempted.

1b Lichens which bear their spores in sacs (asci). Figs. 117 to 126. ...2

ORDER ASCOLICHENES

2a (a, b, c, d) Closely growing scale-like (crustose) lichens (some parasitic species) with very minute projecting fruiting bodies (stipe) which are usually not branched. Algal symbiont, *Chlorococcum*.

Family CALICIACEAE

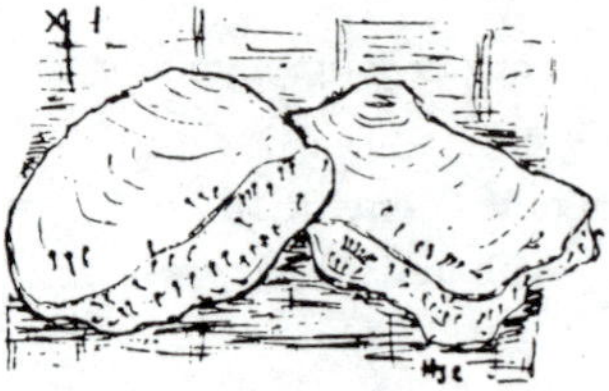

Fig. 117. *Calicium polyporaeum* Nyl.

Grows parasitically on bracket fungi. The thallus so blends with its host as to be almost indistinguishable. They are usually so small as to be difficult to locate, and have not even been thought of as lichens by some botanists.

Figure 117

2b **Apothecia elongate and often branched. Algal symbiont usually *Trentepohlia* (Family Chlorophyceae) Thallus crustose.**

Family GRAPHIDACEAE

Fig. 118. *Graphis scripta* Ach.

Thallus smooth, thin, ashy or olive on bark of trees. Apothecia ⅛ inch or more in length. While many lichens are at their best in the North the members of this family are more abundant farther south.

Figure 118

2c **Apothecia disk or cup-shaped (a few with nearly closed fruiting bodies). Thallus of many forms. Figs. 120 to 126. ....3**

2d **Apothecia surrounded by a perithecium leaving only an ostiole at apex. Symbiont *Trentepohlia* or *Protococcus*. Thallus foliose.**

Family DERMATOCARPACEAE

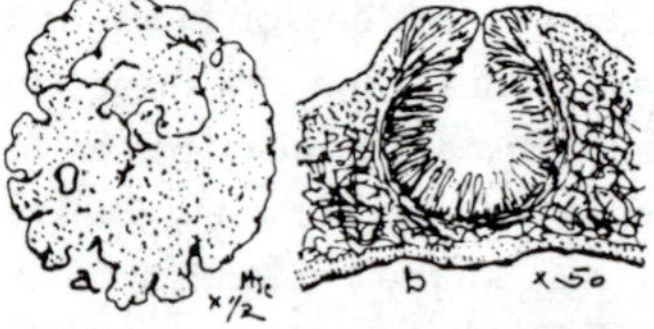

Fig. 119. *Dermatocarpon miniatum* Fr. a, typical thallus; b, section through fruiting body.

Usually found on limestone. It is very widely distributed.

Figure 119

3a **Thallus and apothecia deep yellow or orange; form variable.**

Family TELOSCHISTACEAE

Fig. 120. a, *Placodium elegans* Ach.; *Telochristes chrysopthalmus* Fr.; b, part of thallus; c, ascus and paraphyses.

This family contains a number of very widely distributed species, some being world-wide in their living.

Figure 120

3b Not deep yellow or orange. Figs. 121 to 126. 4

4a (a, b, c) With prominent erect fruiting bodies sometimes branched and often brightly colored. Figs. 122 and 123. 5

4b Main part of plant a leaflike thallus. Figs. 124 to 126. 6

4c Main part of plant crustose. Apothecia rounded.

Family LECIDEACEAE

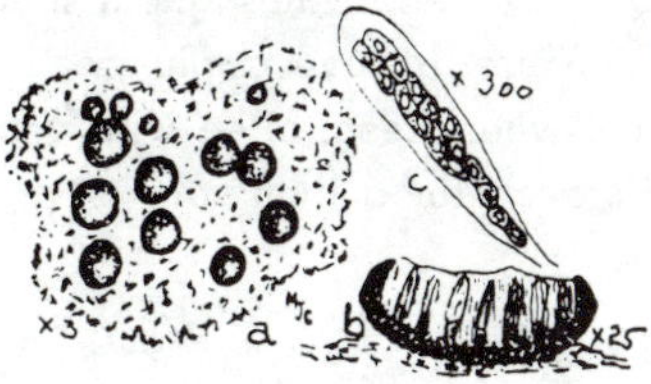

Fig. 121. *Buellia myriocarpa* Mudd., a, typical plant; b, section through fruiting body; c, ascus.

This is a large family with many widely distributed species.

Figure 121

5a With basal vegetative part and erect fruiting bodies (podetia) simple or somewhat branched. Fruiting bodies often scarlet or brown with concave or convex hymenia. **Family CLADONIACEAE**

Fig. 122. a, *Cladonia rangiferina* Web., Reindeer Moss. A beautiful ashy-gray plant often growing in great abundance over earth and rocks. Widely distributed.

b, *Cladonia pyxidata* Hoff.

On earth and rotten wood; cosmopolitan.

Figure 122

5b Basal thallus disappearing before maturity. Apothecia at ends of much branched fruiting bodies. **Family STEREOCAULACEAE**

Fig. 123. *Stereocaulon coralloides* Fr.; a, typical plant; b, tip of branch showing small disk-like apothecia.

These plants often grow in dense clusters on the ground or on rocks.

Figure 123

6a (a, b, c) Thallus rather closely attached to supporting body so as to be only sub-foliose, or somewhat crustose. The algal symbiont usually one of the Chlorococcaceae. **Family LECANORACEAE**

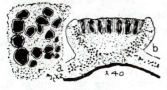

Fig. 124. *Lecanora subfusca* Ach.; a, thallus with apothecia; b, section through apothecium.

Common on trees and sometimes on rock. Light green to whitish with fruiting cups almost black.

Figure 124

43

6b Plants with large foliose parts; apothecia usually buried in the lobes of the thallus, sometimes on under side.

Family PELTIGERACEAE

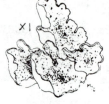

Fig. 125. *Peltigera aphthosa* Willd.

On earth and rocks. Scattered throughout much of the northern hemisphere. Above apple-green to brownish, below white when fresh, often with veins dark. Sometimes mistaken for a liverwort.

Figure 125

6c Thallus plainly foliose or sometimes fruticose.

Family PARMELIACEAE

Fig. 126. *Parmelia perlata* Ach.; a, typical plant; b, ascus and paraphyses.

Common on rocks; widely scattered.

c, *Usnea barbata* Fr., a hanging form which roughly resembles "Spanish Moss".

Figure 126

KEY TO THE FAMILIES OF THE FUNGI

1a Vegetative body, a plasmodium (naked slimy mass of living substance flowing in decaying wood, leaves, etc., usually in the dark). Reproductive bodies, tiny knob or plume-like structures (sporangia) usually in groups and of various colors.
Figs. 138 to 144. (The Slime Molds). 15

1b Not as in 1a. 2

2a Microscopic single celled plants (often clinging in groups or chains). Reproduction by dividing through middle (fission).
Figs. 127 to 137. (The Bacteria). 3

2b Fungi with vegetative body of filaments and various reproductive organs. Figs. 145 to 200. (The True Fungi). 21

THE BACTERIA (SCHIZOMYCETES)

3a Cells living only parasitically within the cells of animals.

Order RICKETTSIALES
Family RICKETTSIACEAE

Dermocentroxenus rickettsii, Rocky Mountain Spotted Fever.
It is transmitted by the bite of some species of ticks which in turn have gotten the organism from some of the rodents which serve as reservoirs for the disease. The disease is now also found in regions other than the Rocky Mountains. No drawing has been attempted since there is so little one can picture for some of the smallest bacteria.

3b Cells capable of living other than as parasites within the cells of
animals. 4

4a Cells self sustaining by oxidation of sulphur or by photosynthesis
through green or purple pigments. Plant-like spores rarely if ever
found. The Sulphur Bacteria.   Order **THIOBACTERIALES**
Cells containing basteriopurpurin, and sometimes sulfur granules.
Cells of various shapes but not filamentous.
Family RHODOBACTERIACEAE

The species of this rather large family may be spherical or rod
shaped and single cells or in colonies. The red or purple color is
characteristic. No picture has been attempted since the color is the
most distinguishing factor.

4b Cells not as in 4a. 5

5a Cells attached by a gelatincus base   or with a gelatinous stalk.
The Stalked Bacteria.   Order **CAULOBACTERIALES**
With stalks of ferric hydroxide.   Family **GALLIONELLACEAE**

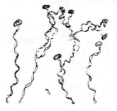

Fig. 127. *Gallionella ferruginea.*
    The attaching bands are ribbon-like and much
twisted.

Figure 127

5b Cells without a gelatinous stalk or base. Figs. 128 to 137. 6

6a Cells collecting into masses and forming cysts or fruiting bodies.
The Slime Bacteria.   Order **MYXOBACTERIALES**

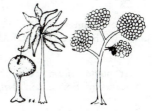

Apparently not very well known. One
family of course would be
MYXOBACTERIACEAE

Fig. 128. Fruiting bodies of three species
of this order.

These plants in some ways, of course,
resemble the Slime Molds.

Figure 128

6b Cells not as in 6a. Figs. 129 to 137. 7

7a Very slender spiral rods, motile but without apparent flagella.
Protozoa-like. Order SPIROCHAETALES
Family SPIROCHAETACEAE

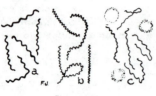

Figure 129

Fig. 129. a, *Treponema pallidum.*
This organism is the causative agent of syphilis in man.
b, *Spirochaete plicatelis;*
c, *Spirochaete obermeieri.*
The spirochetes are slimmer than the members of the genus *Spirillum* and are more like the protozoa.

7b Cells not as in 7a. Figs. 130 to 137. 8

8a Cells (usually filamentous) covered with a gelatinous sheath which is often encrusted with an iron compound. The Sheathed or Iron Bacteria. Order CHLAMYDOBACTERIALES
Family CHLAMYDOBACTERIACEAE

Figure 130

Fig. 130. a, *Crenothrix polyspora;* b, *Leptothrix hyalina.*
These are apparently forms between the true bacteria and the simplest algae and protozoans. They are known as "Higher bacteria" or *Trichobacteria.*

————————*————————

8b Cells not sheathed. Figs. 131 to 137. 9

9a Cells often branched, usually elongated or filament-like and sometimes forming a mycelium. Figs. 136 and 137. 14

9b Cells not forming a mycelium or branching and not especially elongate. The True Bacteria. Figs. 131 to 135. 10

ORDER EUBACTERIALES

10a Cells self sustaining by oxidizing a nitrite or ammonia.
Family NITROBACTERIACEAE

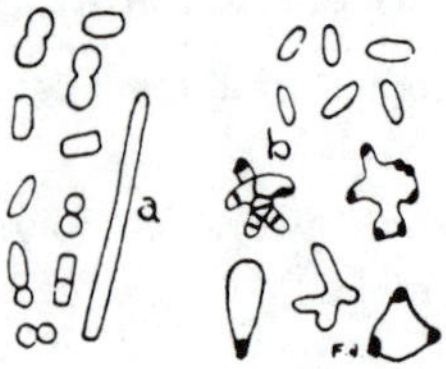

Figure 131

Fig. 131. a, *Acetobacter pasteurinum* Beig.; b, Normal and involute forms of *Rhizobium leguminosarum* Frank.
They play an important part in the activities of soils and sewage disposal.

10b Cells not living on inorganic materials. Figs. 132 to 135. 11

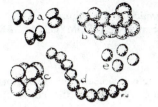

11a Cells spherical in shape although often becoming oval before dividing.

Family COCCACEAE

Fig. 132. a, cells in pairs, *Diploccus*; b, in irregular masses, *Staphylococcus*; c, in cubes, *Sarcina*; d, in chains, *Streptococcus*; e, cells scattered, *Micrococcus*.

Figure 132

11b Cells normally elongated; especially when in active growth. Figs. 133 to 135. 12

12a Cells curved or spiral usually with flagella at the ends.

Family SPIRILLACEAE

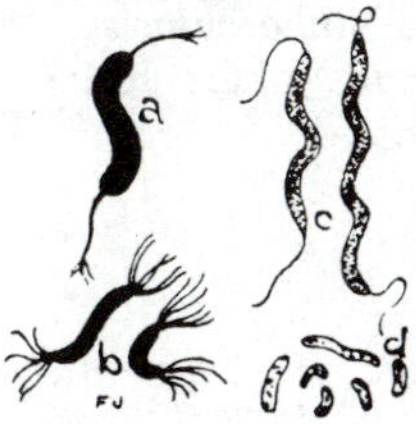

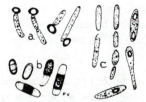

Fig. 133. a, *Spirillum undulata*; b, *Spirillum rubrum*; c, *Spirillum volutans*; d, *Vibrio cholerae*.

In studying movement of bacteria the student should not confuse it with "Brownian movement", the trembling of minute objects when highly magnified. This is likely due to a molecular bombardment.

Figure 133

12b Cells straight or but slightly bent. Figs. 134 and 135. 13
13a Species producing spores within the cells (endospores).

Family BACILLACEAE

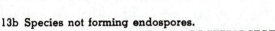

Fig. 134. a, *Bacillus subtilis* Cohn.; b, *Bacillus cohaerens*; c, *Clostridium butyricum* Praz.

Spores offer a means of living through unfavorable times. These bacteria, then, are especially hard to kill.

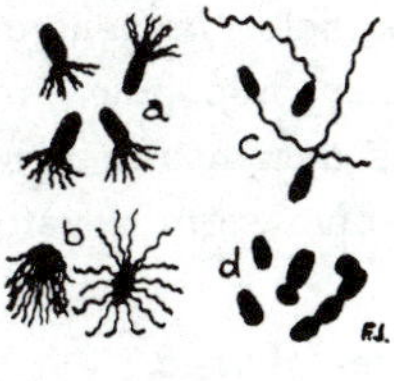

Figure 134

13b Species not forming endospores.

Family BACTERIACEAE

Fig. 135. a, *Pseudomonas syncyanea*; b, *Proteus vulgaris*; c, *Pseudomonas macroselmis*; d, *Escherichia coli*.

Only part of the members of this family are motile. The family ranks at the top when considered from the economic viewpoint.

Figure 135

ORDER ACTINOMYCETALES

14a Filamentous and branching, often forming a mycelium.

Family ACTINOMYCETACEAE

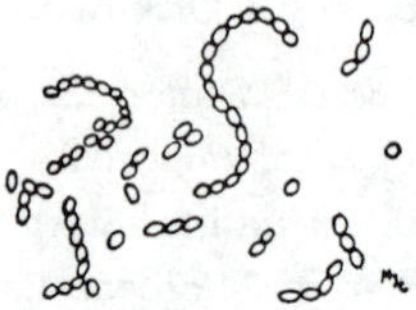

Fig. 136. *Actinomyces* sp.
These bacteria resemble molds in their behavior. One species seems to be related to the cattle disease, lumpy jaw.

Figure 136

14b No mycelium and but rarely branched.

Family MYCOBACTERIACEAE

Figure 137

Fig. 137 a, *Mycobacterium tuberculosis*. The causative agent of human tuberculosis; b, *Corynebacterium diphtheriae*. The diphtheria organism.

FILTERABLE VIRUSES

Some mention needs to be made of the Filterable Viruses even though they do not lend themselves to identification by any of the methods employed here.

Viruses are termed filterable because they pass through porcelain filters which normally "filter out" the smallest bacterial organisms known. Virus particles are ultramicroscopic, that is, too small to be seen by the compound microscope..

The virus of tobacco-mosaic disease, for instance, has been isolated and found to be a high molecular weight crystallizable nucleo-protein. All of the viruses studied have been described as protein-like in character. Several investigators maintain the virus is merely a protein unit, while other workers contend the virus is a living particle. In general, viruses are examples of extreme parasitism in that they propogate only in the presence of living cells. To date, no one has demonstrated the cultivation of a virus on laboratory media. Listed below are several diseases which have a virus as the pathogen:

Influenza, Poliomyelitis, Yellow Fever, Rabies Herpes, Equine Encephalitis and Tomato Bushy Stunt.

THE SLIME MOLDS (MYXOTHALLOPHYTA)

15a Species that live as parasites on other plants. (Body within tissue of host naked, multinucleate and usually not forming a wall even when spores develop.) Order PLASMODIOPHORALES
Family PLASMODIOPHORACEAE

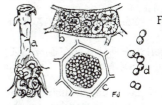

Fig. 138. *Plasmodiphora brassicae*, Club-root of Cabbage. a, stem and root of diseased cabbage plant; b, cell of cabbage filled with amoeboid cells of the parasite; c, cell filled with spores; d, individual spores.

Figure 138

15b Species that are saprophytic. Figs. 139 to 144. 16
16a Vegetative phase of free amoeba which gather in aggregations for fruiting. Found on dung of animals and in soil.

Order ACRASIALES
Family ACRASIACEAE

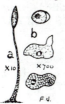

Fig. 139. *Dictyostelium* sp.; a, fruiting body; b, individual cells.

The members of this small family are small and very simple slime molds.

Figure 139

16b Vegetative stage consisting of a mass of slime (plasmodium) which has many nuclei and a flowing movement. Figs. 140 to 144. 17

CLASS MYXOMYCETES

17a Spores borne internally within a sporangium. Fruiting bodies varying widely in shape and color and developing from a plasmodium which has lived within decaying logs, masses of decaying leaves, etc. A large majority of the slime molds belong here. 18

17b Spores borne externally on erect fruiting growths.
Order EXOSPORALES
Family CERATIOMYXACEAE

Fig. 140. *Ceratiomyxa fruticulosa*; a, fruiting pillars; b, tip of pillar enlarged; c, fruiting bodies of another species.

Figure 140

ORDER ENDOSPORALES

18a Sporangia with spores supported by a frame work of netted threads (capillitium). Figs. 141 to 144. 19

18b With no capillitium or very poorly developed. Outer wall of sporangia (peridium) with thinner areas at top.

Family CRIBRARIACEAE

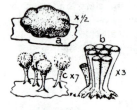

Figure 141

Fig. 141. a, *Enteridium splendens* Morg.; b, *Tribifera ferrugionosa* Macb.; c, *Cribraria argillacea* Pers.

The perfection of these tiny fruiting bodies are a constant source of wonder to the thoughtful student.

19a Deposits of lime plainly evident throughout the fruiting body. Spores black.

Family PHYSARACEAE

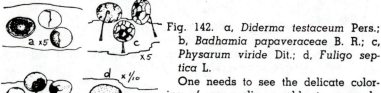

Figure 142

Fig. 142. a, *Diderma testaceum* Pers.; b, *Badhamia papaveraceae* B. R.; c, *Physarum viride* Dit.; d, *Fuligo septica* L.

One needs to see the delicate coloring of many slime molds to properly appreciate them.

19b Limy deposits absent (rarely scanty). Figs. 143 and 144. 20

20a Sporangia (fruiting bodies) with a columella (central shaft) and many capillitium threads (frame work).

Family STEMONITACEAE

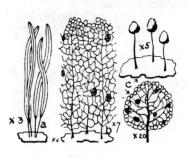

Figure 143

Fig. 143. a, *Stemonitus morgani* Peck; b, *Stemonitus confluens* E. & C.; c, *Comatricha nigra* Pers.

Stemonitus resembles a group of tiny ostrich plumes. The spores of these plants are borne on capillitium threads as shown in b.

20b Without columella; capillitium threads hollow and usually ornamented. Spores usually yellow, never black or purple.

Family TRICHIACEAE

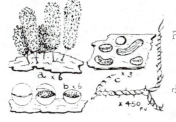

Fig. 144. a, *Arcyria denudata* Pers.; b,
Perichaena corticalis Rost:; c, *Trichia
iowensis* Macb. and capillitium thread.
Many colors, some rather brilliant are
displayed by these fruiting bodies.

Figure 144

THE HIGHER FUNGI (EUMYCETES)

21a Vegetative branches (mycelium) continuous (having no cross walls)
(aseptate). (The Algal-like Fungi). Figs. 145 to 158. 23
21b Mycelium with cross walls (septate) (all higher fungi with fruiting
bodies one-half inch or more across, belong here).
Figs. 159 to 200. 22
22a Spores borne in sacks (asci). The Sac Fungi.
Figs. 159 to 185. 36
22b Spores borne on clubs (basidia). The Club Fungi.
Figs. 186 to 200. 60

THE ALGAL-LIKE FUNGI (PHYCOMYCETES)

23a Sexual reproduction by small spermatazoids and larger eggs
(heterogamous). Sub-class OOMYCETES. Figs. 145 to 153. 24
23b Sexual reproduction by fusion of equal sized motile sex cells
(isogamous). Sub-class ZYGOMYCETES. Figs. 154 to 158. 32
24a Conidia present. Parasitic on other land plants.
Figs. 145 and 146. 25
24b No conidia; reproducing only by sexual spores and zoospores.
Fig. 147 to 153. 26

ORDER PERONOSPORALES

25a Sporangiophores extending beyond the host tissue, usually
branched; conidiospores not in chains.

Family PERONOSPORACEAE

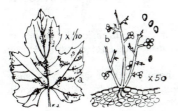

Fig. 145. *Plasmopara viticola* B. & D.,
Downy-mildew of Grapes; a, Grape
leaf with patches of downy mildew; b, conidiophores and spores
arising from tissue of host.

Many disease-producing fungi are
included in these "downy mildews".

Figure 145

25b Sporangiophores forming under the epidermis of the host plant, unbranched; spores in chains.　　　　　**Family ALBUGINACEAE**

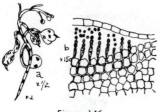

Figure 146

Fig. 146. *Albugo candida* Kuntz., White Rust of Crucifers; a, deformed radish plant infected with *Albugo;* b, Section showing growth of conidia on host.

The white blister-like sori of these plants give the name "white rust". The diseased parts are deformed and swollen.

————————*————————

26a Plant filaments (mycelia) poorly developed; often only a single cell. Parasitic on other fungi, algae or seed bearing plants. Figs. 147 to 149. 27

26b Plant filaments of more than one cell though sometimes much reduced. Figs. 150 to 153. 29

ORDER CHYTRIDALES

27a Reproductive zoospores and gametes without flagella; a globular fertile part outside of host with thread-like vegetative parts extended inward.　　　　　**Family RHIZIDIACEAE**

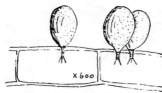

Figure 147

Fig. 147. *Rhizophidium ovatum* Couch. This like many other parasites makes a rapid growth, the life cycle in this case requiring less than one day.

27b Reproductive cells with one or more flagella; living wholly within the host. Figs. 148 and 149. 28

28a All tissue fertile; confined to a single host cell.　　　　　**Family OLPIDIACEAE**

Figure 148

Fig. 148. *Olpidium brassicae* Dang; a, resting spores; b, spores with single flagellum.

Some of these parasites have both summer and winter spores to better fit them for their bandit life.

28b Thread-like parts alternating with swollen parts; wandering through several host cells.   Family CLADOCHYTRIACEAE

Fig. 149. *Cladochytrium replicatum* Karl.
This species grows as a parasite on several species of algae as well as on some aquatic seed-bearing plants.
Other species parasite violets, grasses, grapes, cabbage, etc.

Figure 149

29a Fertilization by motile sperms; zoospores with but one flagellum. Aquatic saprophytes.   Order MONOBLEPHARIDALES
Family MONOBLEPHARIDACEAE

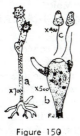

Fig. 150. *Monoblepharis polymorpha* Cor.; a plant with reproductive organs; b, antheridium, oogonium and uniflagalated sperm; c, zoospore.
These saprophytes are found in fresh-water pools.

Figure 150

29b Sperms non-motile; fertilization through an antheridal tube.
Figs. 151 to 153. 30

ORDER SAPROLEGNIALES

30a Vegetative mycelia, thin hyphae of uniform diameter; zoosporangia usually globular and much broader than the mycelium; zoospores biflagellated. Often parasitic on plants.
     Family PYTHIACEAE

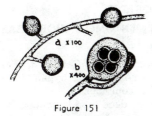

Fig. 151. *Pythium* sp.; a, part of plant with sporangia; b, antheridium and oogonium.
It will be noted that the antheridia (male organs) in these plants as well as some others contact the eggs, making it unnecessary for the sperms to be motile.

Figure 151

30b Vegetative mycelia thick tubular hyphae; aquatic. Hypha not much thickened to form the cylindrical zoosporangia.
Figs. 152 and 153. 31

31a Filaments with regular constrictions; often branched.

Family LEPTOMITACEAE

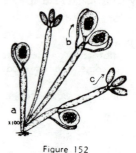

Figure 152

Fig. 152. *Sapromyces andrognyus* Thaxter; a, part of plant with; b, antheridia and oogonia and; c, sporangia.

The members of this family attach themselves to the decaying plant parts on which they live.

31b Filaments of uniform diameter not constricted. Often found growing on dead fish and other animals in water. Water Mold.

Family SAPROLEGNIACEAE

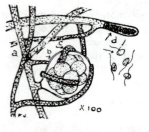

Figure 153

Fig. 153. *Saprolegnia* sp.; a, coenocytic filament; b, oogonium with eggs; c, antheridium; d, sporangium and zoospores.

These plants are known as "water molds". They may be easily produced for study by leaving dead insects in pond water for a few days. The spores of course are already in the water when it is collected. That makes it desirable to get water at several places.

————————*————————

32a Asexual reproduction by conidiospores. Some species parasitic on insects, others saprophytic. Figs. 154 and 155. 33

32b Asexual reproduction by means of aplanospores borne in sporangia. The Black Molds. Figs. 156 to 158. 34

ORDER ENTOMOPHTHORALES

33a Parasitic on spiders and insects. Family ENTOMOPHTHORACEAE

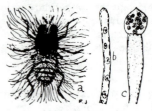

Figure 154

Fig. 154. *Empusa muscae* Cohn:; a, Fly killed by this parasite; b, hypha from body of fly; c, conidiophore.

The flies found dead and attached to window panes, etc., have usually been killed by this fungus. A fly swatter is quicker and more certain, however.

33b Parasitic on plants, or saprophytic.
Family BASIDIOBOLACEAE

Figure 155

Fig. 155. *Basidiobolus ranarum.*

Mycelium with reproductive organs. This species grows on the dung of frogs.

A few species live on higher fungi and one species is parasitic on the fern prothalli.

ORDER MUCORALES

34a Asexual spores borne in typical sporangia. Figs. 157 and 158. . .35

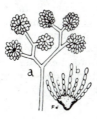

Figure 156

34b Asexual spores conidia-like, sometimes accompanied by larger sporangia with many spores. Conidiospores in chains.
Family PIPTOCEPHALIDACEAE

Fig. 156. *Piptocephalis* sp.; a, conidiophore; b, conidial head with spores.

This fungus is a saprophyte and lives among other fungi. The chains of conidia-like spores are characteristic.

35a Sporangium with a columella; zygospore but thinly covered or naked. A large family including the bread molds and many others.
Family MUCORACEAE

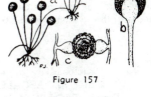

Figure 157

Fig. 157. *Rhizopus nigricans* Ehr.; a, showing habits of growth; b, sporangium with columella; c, zygospore.

This is the mold that greets you when you open the bread box. Two strains are required if zygospores are to be produced.

35b Sporangium without a columella; zygospore surrounded by a thick wall of the hypha. **Family MORTIERELLACEAE**

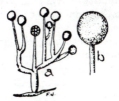

Figure 158

Fig. 158. *Mortierella candelabrum;* a, branched sporangiophore; b, sporangium.

In this family both sporangiophores and conidia are often produced on the same plant.

The molds are highly destructive but play an important role in reducing unwanted organic matter.

THE SAC FUNGI (ASCOMYCETES)

36a Without definite fruiting bodies; asci formed directly from the zygote and borne singly on a mycelium. Sub-class PROTOASCO-MYCETAE. Figs. 159 and 160. 37

36b Asci contained in a definite fruiting body. Asci arising indirectly from the zygote. Figs. 161 to 185. Sub-class EUASCOMYCETAE..38

ORDER SACCHAROMYCETALES

37a Asci similar to vegetative cells. Vegetative cells solitary or but loosely attached in strands scarcely forming a mycelium.
The Yeasts.   Family SACCHAROMYCETACEAE

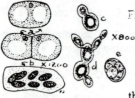

Fig. 159. Fission Yeast, *Schizosaccharomyces octosporus* Bey.; a, vegetative cells; b, an ascus with ascospores; Budding yeast, *Saccharomyces cerevisiae* Han., c, budding cell; d, chain; e, ascus with ascospores.

Yeasts function not only in bread-making and the production of alcohol but also cause many food products to ferment and decay.

Figure 159

37b Asci terminal or intercalary on definite mycelia.
                         Family ENDOMYCETACEAE

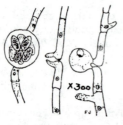

Fig. 160. *Eremascus fertilis* Stop. Stages in reproduction.

Grows as a mold on jellies and other similar food products.

This and other near relatives are particularly interesting in connecting the yeasts with the other Ascomycetes.

Figure 160

38a Asci grouped in a pallisade-like layer but not enclosed by a peridium.
                              Order EXOASCALES
                              Family EXOASCACEAE

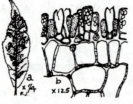

Fig. 161. *Exoascus deformans* Fcl.; a, diseased peach leaf; b, section through leaf showing asci.

Often common on peach trees but does not seem to do a great deal of harm.

Figure 161

38b Asci definitely related to a fruiting body (ascocarp) which may be widely open or nearly enclosed. Figs. 162 to 185. 39

39a Ascocarp (fruiting body) when mature, open and more or less cup-like (apothecium). Figs. 162 to 174. 40

39b Ascocarp when mature, globular or cylindric enclosing the asci (perithecium). Figs. 175 to 185. 49

40a (a, b, c, d) Apothecium disk, saucer or cup-shaped, with the asci standing parallel in a hymenial layer lining the cup; often closed when young; from very minute to 4 or 5 inches in diameter. Figs. 162 to 167. 41

40b Fruiting tissue which is usually borne on a stalk exposed from the first, club-shaped or convex; with pits or ridges. Figs. 168 to 170. 45

40c Fruiting tissue covered by a tough membrane and not becoming exposed until nearly mature. Figs. 171 to 174. 47

40d Subterranean tuber-like fruiting bodies containing chambers in which the asci are borne. These truffles live on the roots of trees.
Order TUBERALES
Family TUBERACEAE

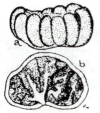

Fig. 162. *Tuber* spp.; a, exterior view; b, cross section.

In Europe where these are much used for food, pigs and dogs are used to locate the truffles for digging.

Figure 162

ORDER PEZIZALES

41a Ascocarps leathery or horny, disk or plate-shaped and free from the first; paraphyses united to form a covering over the asci.
Family PATELLARIACEAE

Fig. 163. *Patella scutellata* Morg.; a, as seen on bark; b, ascocarp in cross section; c, ascus.

On rotten wood or sometimes on soil, vermilion. Widely distributed.

Figure 163

41b Ascocarps waxy, fleshy or gelatinous, ends or paraphyses not uniting. Figs. 164 to 167. 42

42a The outer coat of the ascocarp (peridium) blending into the inner coat (hypothecium). Figs. 165 to 167. 43

42b The peridium and the hypothecium distinct; peridium of thin-walled translucent cells.   **Family HELOTIACEAE**

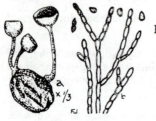

Figure 164

Fig. 164. *Sclerotinia fructicola* Rehm.; American Brown Rot. A very destructive disease of stone fruits; a, apothecia growing on "mummy" plums; producing ascopores, b, conidiophores and conidia as they grow in great abundance on the decaying fruit.

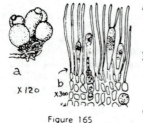

Figure 165

43a Fruiting layer of ascocarp, convex, and open from the first with little or no peridium.   **Family PYRONEMACEAE**

Fig. 165. *Pyronema confluens* Obs.; a, oogonia and antheridia; b, cross section through young apothecium.

Common and widely distributed on charcoal in burned-over ground.

43b Fruiting layer of the ascocarp, concave with a fleshy peridium. Figs. 166 and 167. ....**44**

44a Upper surface of hymenial (fruiting) layer smooth without the asci standing above the rest of the layer. Apothecia either stalked or sessile.   **Family PEZIZACEAE**

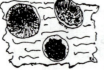

Figure 166

Fig. 166. *Peziza repanda* Pers.

Diameter up to 4 inches, whitish outside, pale brown within.

The members of this large family are often attractively colored.

44b Asci mature projecting above the rest of the hymenial layer. Apothecium without stalk.   **Family ASCOBOLACEAE**

Figure 167

Fig. 167. *Ascobolus magnificus* Dodge.

Diameter 1/5 to 1 inch; brownish when mature.

These plants usually live on the ground and are saphophytic on decaying plants in the soil.

ORDER HELVELLALES

45a Ascocarp with stalk. Figs. 169 to 170. 46

45b Ascocarp without stalk, often attached by several rhizome-like strands of mycelia; fleshy or waxy.　　Family RHIZINACEAE

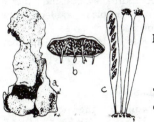

Fig. 168. *Rhizina inflata* Karst.; a, top view; b, lower view; c, asci.

Diameter up to one inch; brown.

These plants often damage forest trees. They are frequently abundant in burned-over areas.

Figure 168

46a Fertile part of the fleshy fruiting body an enlarged head, yellow, green or black; asci opening by a slit or pore. The Earth Tongues.　　Family GEOGLOSSACEAE

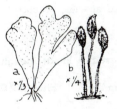

Fig. 169. a, *Mitrula vitellina* Pk. In mossy places. Height to 2 inches, pale yellow. b, *Geoglossum hirsutum*, 2 to 3 inches high, black, surface covered with hairs.

Figure 169

46b Fruiting body a fleshy roughened head growing on a hollow stalk. Asci club-shaped and opening at the end by a cap.　　Family HELVELLACEAE

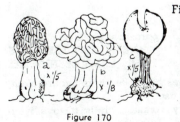

Fig. 170. a, *Morchella esculenta* Pers, Common Morel, 1 to 4 inches high, fruits in the spring, edible; b, *Gyromitra esculenta* Fr., Edible Gyromitra, stem whitish, cap dark red, 2 to 5 inches high; c, *Helvella lacunosa* Aiz., Black-capped Helvella, June to October, 1 to 3 inches high, black, edible.

Figure 170

47a Fruiting ascocarp elongate, opening by a fissure running lengthwise. Figs. 172 to 174. 48

47b Fruiting ascocarp rounded, opening by radiating or star-shaped fissures. **Order PHACIDIALES**

Ascocarps leathery, black and remaining sunken in the tissue of the host. **Family PHACIDIACEAE**

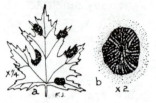

Fig. 171. *Rhytisma acerinum* Fr., Leaf-blotch. A parasite causing black leaf-spots on maples; a, leaf with spots; b, ascocarp.

The family is a large one, but not many species are seriously harmful.

Figure 171

ORDER HYSTERIALES

48a Ascocarp covered by other tissue, its walls attached to its covering. **Family HYPODERMATACEAE**

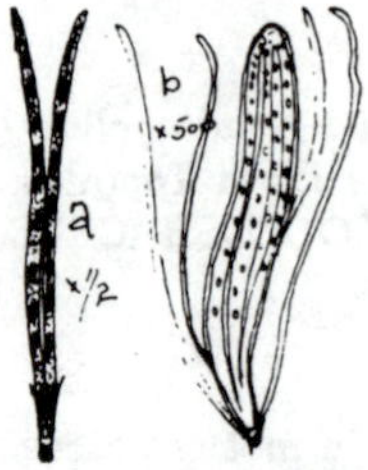

Fig. 172. *Lophodermium pinastri* Chev., produces a blight on leaves of the Scotch pine; a, leaves of pine showing blight spots; b, ascus and paraphyses.

There are around a hundred species, practically all of them saprophytes.

Figure 172

48b Ascocarps free from the first; walls black; usually linear though sometimes round, oval or shield-shaped.

Family HYSTERIACEAE

Fig. 173. *Hysterographium froxini* DeN. Common and widely distributed on Ash; a, twig of host with fruiting bodies; b, ascocarps, enlarged; c, ascus with ascospores.

This family contains a few hundred species which include several parasites.

Figure 173

49a Microscopic fungi living on insects as parasites; perithecia stalked. **Order LABOULBENIALES**
Family LABOULBENIACEAE

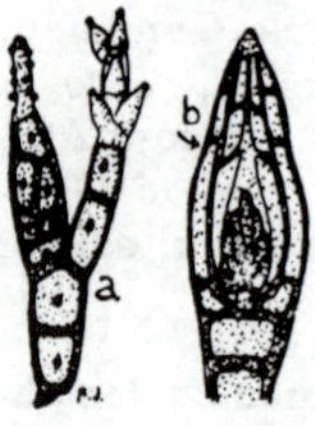

Fig. 174. *Stigmatomyces baerii* Peyri.; a, mature plant; b, perithecium with asci.

Apparently not very common. It is always interesting to see the plant turn the tables and live at the expense of some insect.

Figure 174

49b Perithecium not stalked, solitary and free or united and enclosed in a supporting body. Figs. 175 to 185. 50

50a Asci arranged irregularly within the perithecium.

Order ASPERGILLALES

Ascocarps mostly sessile; and never submerged; mostly saprophytic and spreading asexually by conidiospores which are produced in extraordinary abundance. Many colors and shades.

Family ASPERGILLACEAE

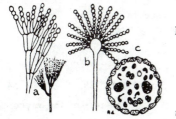

Fig. 175. *Penicillium commune;* a, conidial head showing spore formation; *Aspergillus* sp.; b, typical conidial head; c, section through ascocarp showing asci.

This is the family of *Penicillium notatum* from which penicillin is made.

Figure 175

50b Asci forming at a uniform level. Figs. 176 to 185. 51

51a Perithecium globular, scattered and with no opening. External parasites. Figs. 176 and 177. 52

51b Perithecium with opening. Figs. 178 to 185. 53

ORDER ERYSIPHALES

52a Mycelium white; appendages extending from perithecium. Powdery Mildews. Family ERYSIPHACEAE

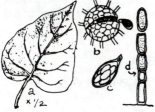

Fig. 176. *Microsphaera alni,* Powdery Mildew of Lilac; a, leaf of Lilac showing the mildew; b, perithecium showing asci; c, an ascus; d, conidiophore.

Powdery mildews are exceedingly common, particularly in damp seasons.

Figure 176

52b Outer mycelia dark colored. Perithecia without true appendages. Family PERISPORIACEAE

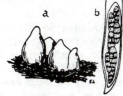

Fig. 177. *Apiosporium salicinum* Kze.; a, plant with perithecia; b, ascus.

Grows on the leaves of trees and shrubs.

The members of this large family are mostly parasites.

Figure 177

53a Perithecium bright colored, fleshy or membranous.

Order HYPOCREALES
Family HYPOCREACEAE

Fig. 178. *Claviceps purpurea* Tul.; a, head of rye with ergoted grains; b, diseased grain (sclerotium) developing ascocarps; c, section through ascocarp, showing asci.

The drug ergot is used in medicine. This plant attacks many species of grasses.

Figure 178

———————*———————

53b Perithecium dark colored and hard. Figs. 179 to 185. ....**54**

54a Perithecium dark colored and distinct from the rest of the mycelium. Figs. 180 to 185. ....**55**

54b Perithecia imbedded in the mycelium and not distinct from it.

Order DOTHIDIALES
Family DOTHIDIACEAE

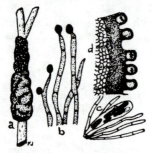

Fig. 179. *Plowrightia morbosa* Sac., Black-knot of Cherry; a, twig with diseased growth; b, conidia; c, ascus; d, section bearing perithecia.

This disease attacks both our cultivated and wild cherries and many of our plums.

Figure 179

ORDER SPHAERIALES

55a Perithecia borne within a stroma (special supporting body). Figs. 184 and 185. ....**59**

55b Plants without a stroma. Figs. 180 to 183. ....**56**

56a Perithecia free upon the substratum or enclosed at their base by a mycelial growth. Family SPHAERIACEAE

Fig. 180. *Rosellinia radiceperda* Mas.; a, fungus on host (cabbage); b, perithecium enlarged; c, conidiophore; *Trichosphaeria sacchari* Mas.; d, fungus on sugar cane; e, perithecia enlarged; f, conidiophores and spores.

Figure 180

56b Perithecium usually wholly covered by the mycelium or epidermis of the host. **57**

57a Perithecium usually with a beak; its wall tough and leathery; asci thickened at apex and opening through a pore.

Family GNOMONIACEAE

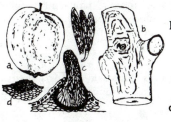

Fig. 181. *Glomerella rufomaculans* S. & V., Bitter-rot of Apple and other fruits; a, diseased apple; b, damage to a twig; c, perithecium with asci; d, conidial stage.

This fungus does serious damage to apples, grape, cotton, and other plants.

Figure 181

57b Perithecium without a distinct beak; not darkened with carbon. . .**58**

58a Asci attached to walls of the perithecium; joined together in bunches but with no paraphyses. Family MYCOSPHAERELLACEAE

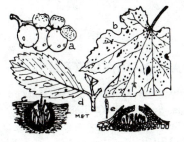

Fig. 182. *Guignardia bidwellii* V. & R., Black-rot of Grapes; a, diseased fruit; b, diseased leaf; c, perithecium with asci (from diseased fruit);
Mycosphaerella fragariae Lind., Leaf-spot of Strawberry, d, diseased leaf; e, perithecium on the host leaf.

Figure 182

58b Asci attached singly to the base of the perithecium; surrounded by paraphyses. Family PLEOSPORACEAE

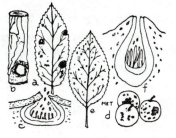

Fig. 183. *Physalospora malorum* Shear., Black-rot of Apple and other Pome fruits; a, leaf with diseased spots; b, effect on twigs; c, section through perithecium; *Venturia pomi* Wint., Apple Scab; d, scab on apples; e, disease spots on a leaf; f, perithecium on leaf.

Figure 183

59a Host and parasite tissues intermingled in the stroma; conidiaspores borne in pycnidia.   Family **VALSACEAE**

Figure 184

Fig. 184. *Valsa* sp.; a, fruiting bodies on limb of host; b, perithecium; c, asci.

More than a thousand species are known, most of which are saprophytes.

59b Stroma of fungus tissue only. Spores blackish, almost always one-celled.   Family **XYLARIACEAE**

Figure 185

Fig. 185. *Xylaria* sp.; a, Fruiting body (stroma); b, cross section through stroma showing perithecia; *Daldinia* sp.; c, stroma on a diseased branch; d, section through stroma showing perithecia in outer margin.

THE CLUB FUNGI (BASIDIOMYCETES)

60a Basidia four-celled, arising directly from resting spores and not forming a hymenium. (Closely associated in a definite structure). Parasitic. Other types of spores often formed, also. Figs. 186 to 189. (Sub-class **HEMIBASIDII**). 61

60b Basidia developing directly from vegetative cells and growing in groups in a hymenium. A few are parasites; mostly saprophytic. Figs. 190 to 200. (Sub-class **EUBASIDII**). 64

61a Great masses of usually black spores (chlamydospores) produced on the host plant, often on the floral parts, especially the ovaries of the grasses. The chlamydospores germinate to vegetative mycelia usually confined to interior of host tissue. Figs. 186 and 187. (The Smuts). 62

61b Chlamydospores, if present borne on definite stalks; usually absent. Basidia arising from teleutospores. Figs. 188 and 189. (The Rusts). 63

ORDER USTILAGINALES

62a Bacidium (promycelium) with spores arising from the side at or near the cross walls. These spores or sprout cells often form budding chains. **Family USTILAGINACEAE**

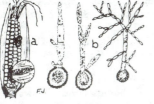

Fig. 186. Corn Smut, *Ustilago zeae* Ung ; a, infested ear of corn; b, basidia with spores and sprout-cells.

Very common in corn fields. May be eaten like mushrooms when young.

Figure 186

62b Basidium (promycelium) with spores clustered at end.

Family TILLETIACEAE

Fig. 187. Stinking Smut, *Tilletia tritici* Wint ; a, clamydospore having germinated and the promycelium produced spores which are conjugating and forming conidia; b, infected head of wheat. Common on wheat.

Figure 187

ORDER UREDINALES

63a Teliospores without stalks; 1 to 4 celled.

Family MELAMPSORACEAE

Fig. 188. *Cronartium ribicola* Fischer, White Pine Blister Rust; a, Gooseberry leaf and teliospores; b, Gooseberry leaf and urediniospores; c, stem of White Pine and aeciospores.

Very destructive to our White Pine forests.

Figure 188

63b Teliospores borne on a simple or compound stalk.

Family PUCCINIACEAE

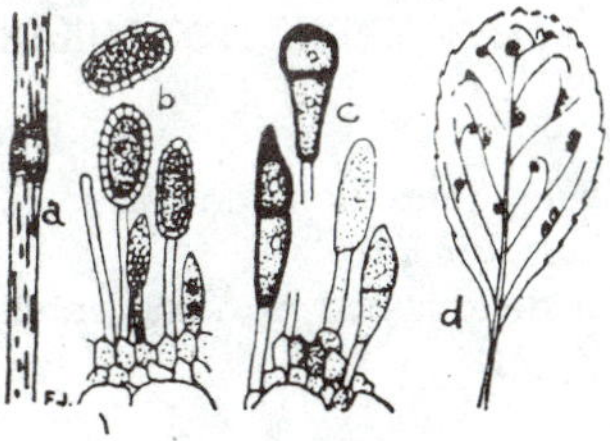

Fig. 189. *Pussinia graminis* Pers., Black Stem Rust of Wheat; a, infected wheat stem; b, uredinospores; c, teliospores; d, barberry leaf with rust spots.

This fungus disease costs immense sums some years to wheat producers. Barberry eradication has helped to control it.

Figure 189

64a Hymenium enclosed within the fruiting body (sporocarp) until they are mature. Figs. 198 to 200. 71
64b Hymenium exposed on some surface part of the fruiting body. Figs. 190 to 197. 65
65a Saprophytes with thin expanded fruiting bodies, gelatinous when fresh or wet, and thorny or leathery when dry. Basidia forked or divided into four cells. Figs. 190 and 191. 66
65b Fruiting body of many forms, often umbrella-like or shelf-like; basidia club-shaped or somewhat cylindrical. Figs. 192 to 197...67

ORDER TREMELLALES

66a Fruiting body shaped somewhat like a human ear. Grows on decaying wood. Basidia divided transversely into four cells.
Ear Fungi.   **Family AURICULARIACEAE**

Fig. 190. Jews Ear, *Auricularia auricula-judae* Schr.

Often abundant in the fall. When fresh, the fruiting bodies are soft and almost jelly-like.

Figure 190

66b Fruiting body irregular, trembling when moist; tough and horny when dry. The Trembling Fungi.   **Family TREMELLACEAE**

Fig. 191. *Tremella* sp.

The jelly-like rather irregular masses constituting the fruiting bodies of this plant are of several colors but always somewhat translucent.

Figure 191

ORDER HYMENOMYCETALES

67a Hymenium (fruiting layer which bears the basidia) smooth. Figs. 192 and 193. 68
67b Hymenium covering projections, folds or pits. Figs. 194 to 197. 69
68a Sporophore (fruiting body) leathery, often small and indefinite in shape.   **Family THELEPHORACEAE**

Fig. 192. a, *Thelophora laciniata* Pers.; b, *Stereum frustulosum* Fri.

Grows on the dead wood of oaks which it disfigures.

Figure 192

68b Sporophore a single club-like part or much divided and branched, usually fleshy; often resembling coral; delicately tinted and colored. Coral Fungi.   **Family CLAVARIACEAE**

Fig. 193. a, *Clavaria formosa* Pers.; b, *Clavaria pistillaris* F., Indian Club Clavaria.
Most of the members of this family are edible. The fruiting bodies are often large and really beautiful when fresh.

Figure 193

69a Spore-bearing hymenium on suspended teeth or similar projections. Teeth Fungi.   **Family HYDNACEAE**

Fig. 194. a, *Hydnum repandum* Fr.; b, *Hydnum coralloides* **Fr.**
Both of these species are edible.
Several hundred species are known. Some are leathery in texture.

Figure 194

69b Spore-bearing hymenium covering the surface of gills (plate-like structures hanging from under side).   **Family AGARICACEAE**

Fig. 195. a, *Pholiota praecox* Pers.; b, Oyster Mushroom, *Pleurotus ostreatus* Fr.
Both of these species are edible. But if you are not sure of your identifications, you'd better be careful. Death is so permanent.

Figure 195

69c Hymenial layer forming the inner wall of pores.
Figs. 196 and 197. 70

70a Fruiting body fleshy; pores readily separating from their support.   **Family BOLETACEAE**

Fig. 196. *Boletus subaureus* Pk.
Yellow with blotches of red-brown; up to 4 inches in diameter. A valuable food species. Grows from early to late fall.

Figure 196

70b Fruiting body woody or leathery; pores not easily separating from supporting tissue. Family POLYPORACEAE

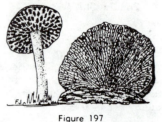

Fig. 197. a, *Polyporus arcularis* Fr.; b, *Lenzites betulina* Fr., showing under surface.

This family is a very large one and common in the forest.

Figure 197

ORDER GASTEROMYCETALES

71a Spore-bearing tissue enclosed in a membranous egg-like structure at first, but later breaking out and growing to some height. Spores adhesive. Plants with strong odor of carrion. Figs. 199 and 200. 72

71b Spores remaining within the fruiting body until mature. Spores dry. The Puffballs, etc. Family LYCOPERDACEAE

Fig. 198. a, *Lycoperdon gemmatus* Bat.; b, *Cythus striatus*, Bird-nest fungus; c, *Geaster hygrometricus* Pers., Earth-Star; d, *Calvatia gigantea* Batsch.

As the species name indicates, this is a large one. Specimens 20 inches in diameter and weighing almost 20 lbs. have been found. All puffballs are edible.

Figure 198

72a Spores borne in a sticky mass (gleba) on the top of the stipe. The Stink Horn Fungi. Family PHALLACEAE

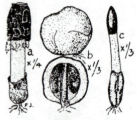

Fig. 199. *Ithyphallus impudicans* Fr., a, mature fruiting body; b, "eggs"; c, *Cynophallus caninus* Fr., Pink-capped Stinkhorn.

These have a way of making their whereabouts known. The flesh flies help spread the spores.

Figure 199

72b Spores borne within the receptacle which is split into a latticed design. Family CLATHRACEAE

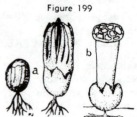

Fig. 200. a, *Anthurus brownii*; b, *Simblum rubescens*.

The spores are borne within the fruiting body rather than on the outside as in the preceding family.

Figure 200

THE BRYOPHYTA (LIVERWORTS AND MOSSES)

The keys for this entire section dealing with the Liverworts and Mosses have been made by Doctor H. S. Conard, author of "How to Know the Mosses".

1a Plants growing flat, scale-like or ribbon-like, usually fork-branched, without distinction of stem and leaf; green or purplish. Figs. 202 to 204 and 206 to 212. Class HEPATICAE (in part). 2

1b Plants with stem and leaves; erect, ascending, prostrate, or hanging from trees. Figs. 201 and 213 to 224. 3

2a Plant opaque by reason of air-spaces inside it; often showing air pores and polygonal markings. Rhizoids with pegs on the inside of the walls. Figs. 204 to 206. 5

2b Plant translucent, watery-looking, without inner air-spaces. Rhizoids without pegs. Figs. 202, 203 and 207 to 212. Class HEPATICAE (in part). 4

3a Leaves in two rows near upper side of stem, without midrib, and with cells isodiametric. Leaves very often notched at apex, or lobed, sometimes with a smaller lobe folded against a larger one. Sporophyte short-lived, the capsule raised on a stalk, splitting into four lobes, emitting spores and slender elaters with spiral bands. Order JUNGERMANNIALES ACROGYNAE. Figs. 201 and 213 to 224. 10

Figure 201

3b Leaves equally spaced all around the stem, usually with midrib; or in two opposite rows, with or without midrib; margine entire or toothed, never notched at apex or lobed; cells elongate to isodiametric. Sporophyte persisting for weeks or months. No elators. Figs. 225 to 253. Class MUSCI. ....page 76

4a (a, b, c) Small rosettes of scales, with surface covered with pear-shaped sacs (involucres) each containing a capsule. No elaters.
Order SPHAEROCARPALES
Family SPHAEROCARPACEAE

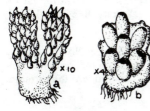

Figure 202

Fig. 202. *Sphaerocarpus texanus* Aust., a, male plant; b, female plant.

The plants of this family and order are quite small and are found on damp ground. The species name of the example used here was given because the type specimen came from Texas. Species names usually have some such significance.

4b Larger (1 cm. or longer at maturity). Spores in a long rod-like capsule which splits in two above as it grows from the base, emitting spores and irregular elaters. No midrib and no gemmae, but sometimes the plant is rough. One chloroplast to each cell.

Order ANTHOCEROTALES
Family ANTHOCEROTACEAE

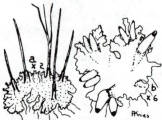

Fig. 203. a, *Anthoceros punctatus*, E. B.; b, *Notothylus orbicularis* (Schwein.)

These plants are known as the Horned Liverworts since the sporophyte has a horned shape. The thallus is often an inch or more in length.

Figure 203

4c Spores in an oval or globular capsule on a slender watery stalk. Capsule splitting into four lobes, emitting spores and spiral-banded elaters. Midrib or mid-furrow distinct. Chloroplasts numerous in each cell. Figs. 207 to 212. .... **6**

ORDER MARCHANTIALES

5a (a, b, c) Air pores visible without a lens, each in a polygonal area. Capsules borne on the under side of an umbrella-shaped cap, with spirally banded elaters among the spores. Capsule wall cells with ring-shaped thickening.   Family MARCHANTIACEAE

Fig. 204. *Marchantia polymorpha* L. a, female plant; b, male plant; *Lunularia cruciata* (L); c, thallus with gemma cups; d, fruiting thallus.

The gemma cups with their asexual gemma-buds offer a unique scheme of multiplication. These tiny bits of the plant grow readily into a new thallus. The gemma cups of *Lunularia* are half-moon shaped.

Figure 204

5b Air-pores not visible without a strong lens. Plants on moist or dry rocks and banks, rarely if ever in neat rosettes. Capsules as above, but with no ring-like thickenings.   Family REBOULIACEAE

Fig. 205. *Reboulia hemisphaerica* G. L. & N.

This little liverwort seems to be scattered rather worldwide. It is found on rocks, walls and on the ground. It is widely distributed.

Figure 205

5c Air-pores, if any, not visible with a hand lens. Plants submerged or floating, or in circular rosettes on very wet ground. Capsules imbedded in the plant, with no elaters among the rough spores.
Family RICCIACEAE

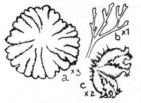

Fig. 206. a, *Riccia frostii* Aust.; b, *Riccia fluitans* L.; c, *Ricciocarpus natans* (L.)

The members of this family are more simple than *Marchantia* but highly interesting. They are found in water and on damp ground.

Figure 206

ORDER METZGERIALES

6a (a, b, c) Plant deeply cut on both sides of a stem-like midrib into wrinkled leaf-like lobes. **Family FOSSOMBRONIACEAE**

Fig. 207. *Fossombronia pusilla* Dum.; a, plant; b, sporophyte; c, spores.

It grows along paths, ditches, etc., on moist ground. It is a small species.

Figure 207

6b Plant with shallow marginal lobes, with lumps of blue-green algae embedded here and there, and with bottle-shaped gemma-containers, or tiny star-shaped gemmae. On moist shaded banks.
Family HAPLOLAENACEAE

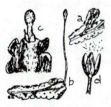

Fig. 208. *Blasia pusilla* L., a, vegetative thallus; b, fruiting thallus; c, sporangium; d, capsule dehiscing.

Grows on damp clay or gravel; green to yellowish-green.

Figure 208

6c Margins of plant even, or wavy but not regularly lobed. The plant itself may be variously lobed or branched. Figs. 209 to 212. 7

7a Midrib well defined, bulging like a cord along lower side of plant, the rest of the plant only one cell thick. Figs. 209 and 210. 8

7b Midrib ill-defined; merely the gradually thickened central part of the plant, which is one cell thick only at the extreme margin if at all. Figs. 211 and 212. 9

8a Plant 1 to 2 mm. wide, much longer than wide, of very even width. Sex organs underneath. On trees, leaves or damp ground.

Family METZGERIACEAE

Fig. 209. *Metzgeria conjugata* Lindb.

Widely scattered. Grows on tree trunks and rocks in shady places.

Several other species are highly similar. The family is not large.

Figure 209

8b Plant 3 to 4 mm. wide, often very irregularly lobed; sex organs on upper side, along midrib. On wet peaty ground.

Family PALLAVICINIACEAE

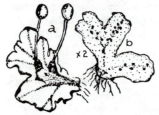

Fig. 210. *Pallavicinia lyellii* Gray, a, fruiting thallus; b, dehiscing capsule; c, antheridium.

Frequently submerged or on wet banks or in swamps. Widely distributed.

Figure 210

9a Plant 4 to 5 mm. wide, usually crowded in wide (10-50 cm.) patches on moist ground. Elaters attached at base of capsule.

Family PELLIACEAE

Fig. 211. *Pellia neesiana* (Gott.). a, fruiting thallus; b, male thallus.

Green tinged with red along veins. On wet ground. Widely distributed.

Figure 211

9b Plant 1 to 5 mm. wide, variously lobed or branched, in shallow water or very wet places. Elaters attached to apex of capsule (tips of valves).

Family ANEURACEAE

Fig. 212. *Riccardia latifrons* Lind., fruiting thallus.

Grows on wet decaying wood.

The family is small but the species are cosmopolitan in their distribution.

Figure 212

ORDER JUNGERMANNIALES

10a Leaves deeply divided into many threads or rows of cells.
Family **PTILIDIACEAE**

Fig. 213. *Ptilidium ciliare* Nees.; a, fruiting plant; b, leaves enlarged; c, involucre with dehiscing capsule.

Rather common; on stumps and rotten logs.

Figure 213

13a (a, b, c, d) Plants brown or purplish, in dense tufts on very wet rocks. Leaves with a rounded notch and two rounded lobes, set transversely on the stem. Perianth shorter than the surrounding leaves and grown fast to them. Family **MARSUPELLACEAE**

Fig. 214. *Marsupella emarginata* Dum.; a, fruiting plant; b, tip of plant with involucral leaves; c, base of pedicel; d, dehiscing capsule.

On wet rock or floating in water. Usually restricted to mountain streams.

Figure 214

13b Plants large (2 to 4 mm. wide) with leaves set very obliquely on the stem, succubous, with margins entire or sharply toothed.
Family **PLAGIOCHILACEAE**

Fig. 215. *Plagiochila interrupta* Dum.; a, fruiting plant; b, leaves in detail.

Occurs on moist banks and rotten logs.

The members of this family are common in our country and in Europe.

Figure 215

13c Leaves entire, or broadly and shallowly notched at apex. Under leaves deeply divided into two long slender sharp lobes (entire in *Harpanshis*). Perianth sharply triangular, with one edge upper.

Family HARPANTHACEAE

Fig. 216. *Chiloscyphus ascendens* H. & W.; a, plant with sporophytes; b, leaves in detail and antheridium; c, involucre; d, ripened capsule.

Found on decaying logs.

Figure 216

13d Leaves entire or 2 or 3 lobed; under leaves entire or absent. Perianth oval or obovoid.

Family JUNGERMANNIACEAE

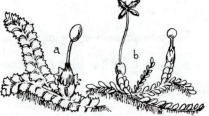

Fig. 217. a, *Jungermannia barbata* Sch:; b, *Nardia crenulata* Lin.

These plants abound throughout the wet areas of the North.

Figure 217

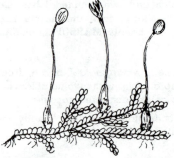
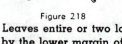

14a Leaves with two or more notches at apex (3 or more lobed); the upper margin of one leaf overlapping the lower margin of the above.

Family LEPIDOZIACEAE

Fig. 218. *Lepidozia reptans* Dum.

Grows on soil and on rotten wood.

The family is a small one but the filmy plants are often very abundant and are found throughout North America and Europe.

Figure 218

14b Leaves entire or two lobed. The upper margin of one leaf covered by the lower margin of the leaf above.

Family CEPHALOZIACEAE

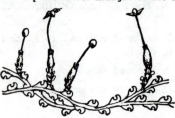

Fig. 219. *Cephalozia multiflora* Spr.

Grows on rotten wood and on the ground.

These plants are very small but are widely distributed.

Figure 219

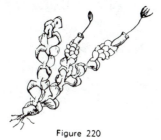

Figure 220

15a Upper lobe of leaf smaller than lower and only partly covering the lower.   Family SCAPANIACEAE

Fig. 220. *Scapania undulata* Dum.

Leaves are reddish or purplish.

Twenty-four species of this genus are known in North America.

15b Upperlobe of leaf much larger than, and completely covering the lower. Figs. 221 to 224. 16

16a (a, b, c) Underlobe of leaf tongue-shaped, attached only at one end; underleaf tongue-shaped, conspiicuous. Large plants; 3 to 8 cm. long.   Family PORELLACEAE

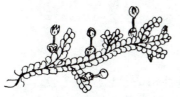

Figure 221

Fig. 221. *Porella platyphylla* Lind.

On rocks and trees in moist places.

This family is better known than some of its near relatives.

16b Underlobe forming a sac or pouch very narrowly attached to upperlobe (or rarely tongue-shaped); underleaves present, notched at apex. Several archegonia in each perithecium. Small blackish or green plants 1mm. wide or less.

Family FRULLANIACEAE

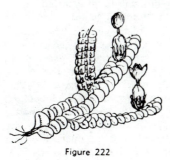

Figure 222

Fig. 222. *Frullania asagraya* Mont.

Common; on rocks and bark of cone-bearing trees.

These plants often grow in large mats. The genus has 26 known species in North America.

16c Underlobe flat, its longest side attached to upper lobe. Figs. 223 and 224. 17

75

17a Underleaves absent; rhizoids attached in tufts to underlobes.
Family RADULACEAE

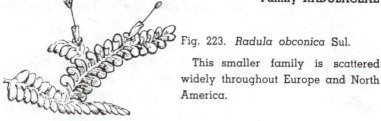

Fig. 223. *Radula obconica* Sul.

This smaller family is scattered widely throughout Europe and North America.

Figure 223

17b Underleaves present, entire or notched; rhizoids in tufts attached to base of underleaf. One archegonium in each perianth.
Family LEJEUNEACEAE

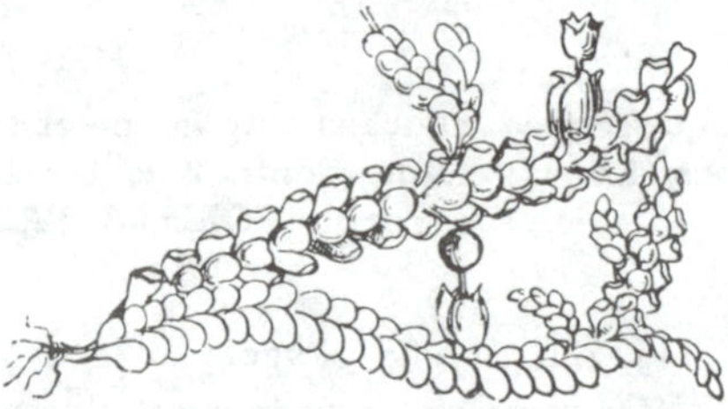

Fig. 224. *Lejeunea clypeata* Sul.; a, leafy plant; b, leaves, top view; c, leaves, lower view; d, sporophyte.
Grows on rocks and trees.
This family reaches its best in the tropics.

Figure 224

THE MOSSES (MUSCI)

1a (a, b, c) Capsule raised on a pseudopodium, spherical, black, shedding a round lid explosively. Leaves whitish, very porous and absorbent. In bogs or wet places. (Peat Moss). Order SPHAGNALES
Family SPHAGNACEAE

Fig. 225. *Sphagnum cymbifolium* Sedw.; a, plant; b, crown of *sphagnum* plant bearing fruit.

This family contains but one genus *Sphagnum*, of which there are many species. The fruit is rather scarce and is peculiar in that its stem is a part of the leafy plant instead of the capsule. This stem is known as a pseudopodium. Peat used as fuel is old, partly decayed sphagnum taken from the bottom of bogs cut into bricks and dried. The dry plants are used for heat insulation and when moistened makes ideal packing for growing plants.

Figure 225

1b Capsule raised on a pseudopodium, cylindric, brownish-black, opening along the middle by four longitudinal slits. Leaves minute, stiff, blackish, with very thick cell walls. On rocks, in mountains. (The Black Mosses)   Order **ANDREAEALES**
     Family **ANDREAEACEAE**

Figure 226

Fig. 226. *Andreaea petrophila* Ehrh.; a, leafy plant with fruit.

All the members of this family are included in the one genus, *Andreaea*. The plants are very small, the slender fragile stems standing usually not over ½ inch tall. The capsule which has no lid divides into 4 vertical parts which diverge when dry.

1c Capsule raised on a rigid seta or nestled among leaves of the plant, dehiscent by a lid, or indehiscent. Of many colors and textures, in all kinds of habitats. Figs. 227 to 253. 2

ORDER BRYALES

2a (a, b, c) Mouth of capsule beset with teeth in a single row, each of which is made up of many cells; teeth without transverse bars or lines. Figs. 228 and 229. 3

2b Mouth of capsule with one of two rows of membranous teeth, or without teeth, or capsules indehiscent. Teeth usually with transverse bars or lines. Figs. 230 to 253. 4

2c Peristome double, the inner being a conspicuous conical plaited membrane, the outer of numerous rod-like rows of cells, or rudiments of these. Capsules oblique and unsymmetric. On banks rich in humus. Saprophytic.   Family **BUXBAUMIACEAE**

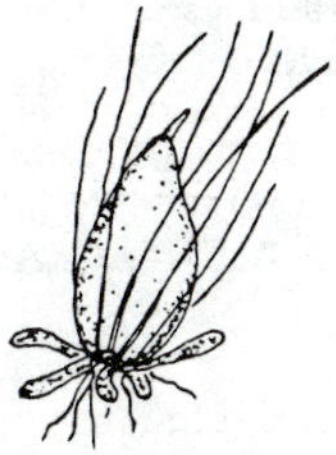

Figure 227

Fig. 227. *Webera sessilis* Lin.

These plants are tiny, have but few or no leaves and in every way are strange-appearing mosses. They are rather widely distributed but rare.

3a Teeth 4, capsule cylindric. Leaves small, ovoid, pointed, with midrib and isodiametric cells. Some stems bear gemma cups at summit. On moist rotten wood, or on moist sandstone.   Family **TETRAPHIDACEAE**

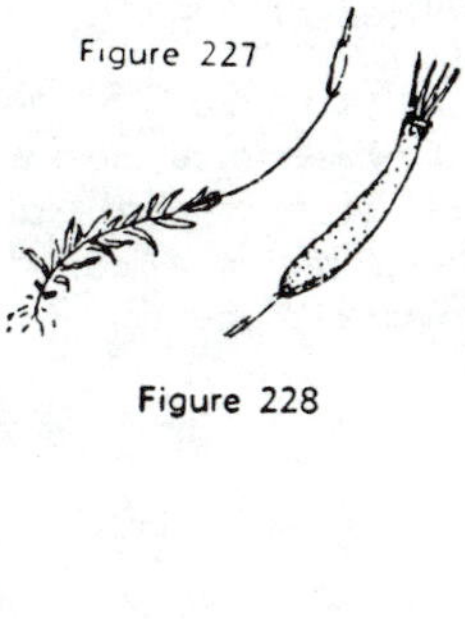

Figure 228

Fig. 228. *Georgia pellucida* Rab.

No other mosses have such small number of peristome teeth. These plants grow in deep shade on wet, rotten wood or sandstone in little tufts. The protonema is scale-shaped.

Figure 229

3b Teeth 32 to 64, their tips attached to a thin membrane covering the mouth of the capsule. Leaves with upright green lamellae along the midrib.

Family POLYTRICHACEAE

Fig. 229. a, *Pogonatum brevicaule* Beauv.; *Polytrichum commune* L. Hair-cap Moss; b, plant with mature capsule; c, sporangium with calyptra.

The plants of this family are large for mosses and are among the most common mosses of many regions.

4a (a, b, c) Peristome consisting of a single row of teeth, each composed (at least at base) of two layers of plates; in the outer layer a single plate forms the width of the tooth; in the inner, two plates go to form the width of the tooth; hence, the tooth seen from *within* shows a fine longitudinal line. If without teeth or indehiscent, the leaf cells are elongate, pointed and smooth, or small, isodiametric and papillose, or small and very thick walled. Fig. 230. Sub-class HAPLOLEPIDEAE......5

Figure 230

4b Peristome characteristically in two circles; an inner thin membrane divided into segments, an outer of 16 (or 8) firm *teeth*. A tooth is composed (at least at base) of two layers of plates; in the outer layer two plates go to form the width of the tooth; the outer surface therefore shows a fine longitudinal line. In the inner layer a single plate forms the width of the tooth. If without teeth or indehiscent the leaf cells are large, isodiametric or sharply rectangular, and smooth walled. If inner peristome is lacking the structure of the teeth will tell. Fig. 231. 11

Figure 231

4c Peristome single, double or none. Leaves tongue-shaped, papillose (hairy). Calyptra cylindric, long pointed above, completely covering the capsule.

Family ENCALYPTACEAE

Figure 232

Fig. 232. *Encalypta streptocarpa* (Hedw.)

The calyptra of these mosses so resembles the extinguisher of a candle that they are sometimes called "Extinguisher Mosses".

5a Peristome teeth 16, split at apex into two prongs; if peristome is absent the leaf cells are long, pointed and smooth, or small and thick walled. Figs. 233 to 253. 6

5b Peristome teeth 16, either undivided or, commonly, diivided into slender threads which are more or less spirally twisted, or imperfect or absent. Upper leaf cells usually papillose, small, isodiametric. Calyptra not covering capsule, early falling off.

Family POTTIACEAE

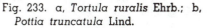

Fig. 233. a, *Tortula ruralis* Ehrb.; b, *Pottia truncatula* Lind.

The peristome teeth in *Tortula* are usually twisted together.

The family has more than a score of known species.

Figure 233

6a Teeth split half way down. Leaves in two opposite rows, each leaf being split at base and standing astride of the next younger leaf (equitant).   Family FISSIDENTACEAE

Fig. 234. *Fissidens adiantoides* **Hedw.**

The leaves are arranged in one plane. These plants reach a height of about ½ inch.

These are aquatic, living in swift moving streams.

Figure 234

6b Leaves spirally arranged, over-lapping like scales of a cone. Figs. 235 to 253. 7

7a Harsh black of blackish-green mosses on dry exposed rocks, sometimes in streams; cell walls thick, often wavy. Leaves often tipped with a colorless bristle. Peristome teeth entire, cleft, perforate or lacking.

Family GRIMMIACEAE

Fig. 235. a, *Grimmia apocarpa* Hedw.; b, *Rhacomitrium aciculare* Brid.

These plants almost always grow on rocks. The leaves often have transparent tips.

This large family contains many common species.

Figure 235

7b Spores small and **very** numerous. Plants and capsules various. Figs. 236 to 253. 8

8a Peristome teeth split nearly to the base into 2 slender strands, or absent. Plants of soil or crevices of rock, medium size to small. Figs. 236 and 237. 9

8b Peristome split half way. Leaves long, slender pointed, erect or curved to one side. Medium sized to large mosses. Figs. 238 to 253. 10

9a Stem distinct, short or long; leaves with midrib. Peristome present or absent. **Family DITRICHACEAE**

Fig. 236. a, *Ditrichum pallidum* Ham.; b, *Pleuridium subulatum* Rab., plant and sporophyte; *Ceratodon purpureus* Brid.; **c**, plant; d, peristome teeth.

Some members of this family attain a height of several inches while others are very small. Decaying wood, rocks and soil are the usual habitat.

Figure 236

9b Stemless; a microscopic cluster of leaves enclosing an indehiscent capsule. Leaf cells rhomboid-hexagonal; midrib present or absent. **Family EPHEMERACEAE**

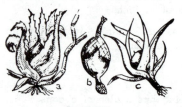

Fig. 237. *Ephemerum serratum* Ham.; a, plant with protonema; b, sporophyte; c, *Nanomitrium* sp., single plant with sporophyte.

The members of this family are very small often less than 1/15 inch high. The protonema often remains active.

Figure 237

10a **Leaves firm without spongy empty cells.** Family DICRANACEAE

Fig. 238. *Dicranum longifolium* Ehr.

Usually yellowish green with long slim leaves. Spores mature in the fall. Very common on rotting logs.

The members of this sizable family are widely distributed.

Figure 238

10b Leaves spongy, with 2 to 4 layers of large empty cells; the chlorophyll cells hidden in the angles. Capsules curved, ribbed, with a swelling on the concave side at base. Plants in dense whitish or pale green cushions. **Family LEUCOBRYACEAE**

Figure 239

Fig. 239. *Leucobryum glaucum* Schimp.

This species is known as the White Moss. Round, thick tufts of these gray-white plants are common in moist woods. The sporophytes are formed sparingly.

11a (a, b, c, d) Inner peristome segments directly in front of outer (not alternating); if peristome is lacking, the leaf cells are rectangular with square ends, smooth. Hypophysis not noticeable. **Family FUNARIACEAE**

Figure 240

Fig. 240. a, *Funaria hygrometrica* Sib. This has been called the Cord Moss. The stem of the sporophyte twists and untwists with changes of moisture; b, *Physcomitrium turbinatum* Brid. This "Common Urn-Moss" is up to ½ inch high.

11b Peristome single, but teeth with 2 rows of plates on outer surface. Hypophysis long or broad or both, often larger than urn, very conspicuous. Stems short, seta long. On dung or decaying vegetable matter at high altitudes or latitudes. **Family SPLACHNACEAE**

Figure 241

Fig. 241. *Tetraplodon bryoides* Lind.; a, typical plant; b, sporangium with calyptra; c, old sporangium; d, peristome teeth.

These mosses are confined in their growth to animal tissues or animal excrement. The greatly enlarged base of the sporangium (hypophysis) is a distinguishing character.

11c Peristome lacking. Capsules nearly erect and globular, small. Leaves in 2 rows, the shoot resembling a microscopic fern. Protonema perennial, in caves, glittering golden green by reflected light. Family **SCHISTOSTEGACEAE**

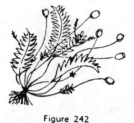

Figure 242

Fig. 242. *Schistostega osmundaceacea* Mohr., typical plant.

This species has been named the Luminous Moss. It grows in dark cavities and caves where is glows with reflected light. The species name refers to its fern-like leaves.

11d Inner peristome segments alternating with the outer. Order EU-BRYALES. Figs. 243 to 253. 12

12a Sporophyte rising from the apex of a main shoot or main branch, with normal leaves around the base. Plants mostly erect. Figs. 243 to 248. 13

12b Sporophyte arising from a peculiar lateral bud whose leaves are very different from ordinary leaves. Plants mostly creeping and branching freely. Figs. 249 to 253. 17

13a (a, b, c) Inner peristome made of slender cilia alternating with 16 short broad outer teeth, the latter often united in 8 pairs. Family **ORTHOTRICHACEAE**

Figure 243

Fig. 243. *Amphidium lapponicum* Sch.; a, plant; b, sporangium with calyptra; c, sporophyte; d, an old capsule.

This is a family of tree-loving mosses. The plants are small and very dark green. The ridges on the dry capsules is a good family character.

13b Inner peristome a circle of prickly cilia. Stout erect mosses. Family **TIMMIACEAE**

Figure 244

Fig. 244. *Timmia* sp., a, plant; b, mature sporangium; c, dry sporangium.

The family is small but the plants are fairly large. There is but the one American genus.

13c Inner peristome a cleft and perforated membrane. 14

14a Leaves papillose. Figs. 245 and 246. 15

14b Leaves smooth, not papillose. Figs. 247 and 248. 16

15a Capsules cylindric, curved, ribbed. Plants often topped by a naked stalk bearing gemma. Leaves broad or narrow, but slenderly tapering.   Family AULACOMNIACEAE

Figure 245

Fig. 245. *Aulacomnium heterostichum* B. & S.; a, plant; b, leaf; c, sporangium and calyptra.

The plants are green, and brownish below. The habitat is moist woods and bogs where it grown on the ground.

15b Capsules globular, ribbed, erect but with lid obliquely placed. Leaves narrow to finely taper-pointed.
                    Family BARTRAMIACEAE

Figure 246

Fig. 246. *Bartramia pomiformis* Hed.; a, plant; b, leaf; c, capsule and calyptra; d, dry capsule.

Medium to large sized plants belong to this family. They grow in sizable tufts in damp shady places.

16a Capsules pear-shaped, horizontal to nodding. Leaves broad to hairlike.   Family BRYACEAE

Figure 247

Fig. 247. *Bryum capillare* L.; a, plant; b, leaf; c, capsule; d, dry capsule.

Fairly large species usually with broad leaves. A large and important family.

16b Capsules barrel-shaped, symmetrical, nodding on a sharply bent seta. Leaves broad, with isodiametric cells. Stems often bending over and rooting at the tip. **Family MNIACEAE**

Fig. 248. *Mnium affine* Bland.; a, male plant; b, female plant; c, leaf; d, dry capsule, and peristome teeth.

The plants of this family are still larger than those of the preceding one.

Figure 248

17a Inner peristome more or less incomplete or even absent; outer teeth with the diplolepideous structure. (See Fig. 231). 18

17b Inner peristome well developed, though cilia may be absent and each segment may be reduced to a double row of plates. Figs. 252 and 253. 20

18a Aquatic; attached to stones or sticks, rarely free. Leaf cells long-linear. Peristome with 16 outer teeth and an inner conical network. Seta short, barely equaling the capsule (Water Mosses). **Family FONTINALACEAE**

Fig. 249. *Fontinalis lescurii* Sul.; a, portion of plant; b, sporophyte; c, mature capsule showing peristome; d, leaf.

Purely aquatic, attached at base with long stems floating.

Figure 249

18b On trees or rocks, never in water. 19

19a Leaves spirally arranged all round the stem; capsule without inner peristome. **Family LEUCODONTACEAE**

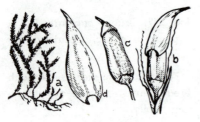

Fig. 250. *Forsstraemia trichomitria* Lin., a, part of plant; b, capsule with calyptra; c, capsult with operculum; d, leaf.

The members of this family are almost wholly confined to trees.

Figure 250

19b Leaves in 2 opposite rows, the twigs therefore flat. Seta short. Capsule erect and symmetric or nearly so.

Family NECKERACEAE

Figure 251

Fig. 251. *Neckera* sp.; a, fruiting plant; b, leaf. Habitat; rocks and trees.

There are both eastern and western species.

20a Leaves with papillae over the cell-cavity. Cells usually minute and indistinct.

Family LESKEACEAE

Figure 252

Fig. 252. *Leskea polycarpa* Ehrh.; a, plant; b, leaf; c, fresh capsule with calyptra; d, dry capsule.

A large family whose species vary rather widely in habitat, size, and form.

20b Leaves smooth, or papillose by projecting ends of cell walls. Cells usually distinct, often very long and slender.

Family HYPNACEAE

Figure 253

Fig. 253. *Brachythecium oxycladon* J. & S.; a, plant; b, leaf; c, capsule; d, dry capsule.

Plants creeping on decaying wood, etc., often fern-like. This genus contains many species of very common and widely distributed mosses. The family is an important one.

85

THE PTERIDOPHYTA (FERNS)

1a Rush-like plants with the cylindrical stems jointed; their nodes collared, with toothed sheaths. Spores borne in terminal cone-like bodies.

Order EQUISETALES
Horsetail Family, EQUISETACEAE

Figure 254

Fig. 254. *Equisetum arvense* L. Field Horsetail. a, fertile shoot; appears in early spring, pinkish brown, 5 to 10 inches high; b, green vegetative shoot; appears after fertile shoot, much branched, sometimes reaching 2 ft. in height; c, *Equisetum robustum* A. Br., Stout Scouring-rush.

The stems of these interesting plants are hollow. They are so heavily impregnated with silica that they were once used for polishing. One tropical species attains a height of 25 ft. Less than fifty species are known.

1b Stems, if any, neither collared nor conspicuously jointed. 2

ORDER FILICALES

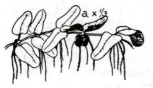

Figure 255

2a Small floating plants with two-ranked flat leaves. Large megaspores and small microspores borne in separate bulb-like fruiting bodies clustered on underside of stem.

Salvinia Family SALVINIACEAE

Fig. 255. *Salvina natans* (L.), Floating Moss; a, plant showing roots and leaves and fruiting bodies.

2b Terrestrial or partly submerged plants but not floating. 3

3a Ferns with distinct tree-like trunks. In greenhouse or tropics.

Tree-Fern Family, CYATHEACEAE

Figure 256

Fig. 256. *Dicksonia antarctica* Lab.

A native of Australia; attains a height of 30 feet with leaves 6 feet or more in length.

Tree-ferns are mostly natives of the tropics or the southern hemisphere. They are palm-like in general appearance but bear spores on the underside or edges of the leaves. Some species attain a height of 50 feet or more. Some 300 species are known.

3b Not tree-like. **4**

4a Plants with creeping stems rooted in mud; leaves resembling 4 leaf clovers. (The Pepperworts).

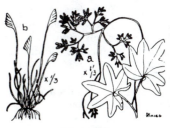

Water-Fern Family, MARSILEACEAE

Fig. 257. *Marsilea quadrifolia* L., European Pepperwort.

In shallow lakes or streams 2" to 10" high. The ovoid fruiting bodies grow from the submerged stem and produce both microspores and megaspores.

Figure 257

4b Not habitually growing in water. **5**

5a Fruiting bodies borne on the back (Fig. 259a) or margins (Figs. 259b and c) of the vegetative leaves or on specialized fertile fronds (Fig. 257). Spores all alike. **6**

5b Spores borne in the axils of narrow, usually imbricated leaves. Stems thickly clothed with small moss-like leaves or leaves quill-like. (The Club Mosses and Quillworts). **9**

6a Twining vine-like ferns or small grass-like plants with spores borne in narrow pinnate spikes.

Climbing Fern Family, SCHIZAEACEAE

Fig. 258. a, *Lygodium palmatum* (Bernh.), Climbing Fern, found in Eastern U. S.; b, *Schizaea pusilla* Pursh., Curly-grass.

Most of the members of this family are tropical, some of which are found in greenhouses.

Figure 258

6b Not as in 6a. **7**

7a Young leaves coiled, Fig. 259. ....**8**

7b Young leaves not coiled; fruiting bodies in a spike or panicle, the sporangia opening by a transverse slit.

Figure 259

Adder's Tongue Family, OPHIOGLOSSACEAE

Fig. 260. a, *Ophioglossum vulgatum* L., Adder's Tongue; b, *Botrychium obliquum* Muhl., Grape Fern.

These simpler ferns are rarer than many others. They should be protected wherever found.

Figure 260

87

8a Sporangia opening vertically (Fig. 261c).

Flowering Fern Family, OSMUNDACEAE

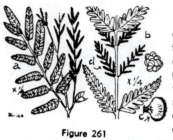

Figure 261

Fig. 261. a, *Osmunda regalis* L., Royal Fern; b, *Osmunda claytoniana* L., Interrupted Fern.

Both of these ferns sometimes attain a height of 6 feet and are widely distributed. The Interrupted Fern is so called because of the fertile pinnae. (Fig. 261d). These become brown and withered presently. We have had these sent in by persons who thought this was a fungus disease and who were seeking a remedy. Of course the prescription must be "Raise some other species of fern".

8b Sporangia opening transversely. The sporangia are stalked and have a raised vertical ring (annulus). They are borne in clusters (sori) on the back of the leaves (sometimes at the leaf margin). Almost all of our common erect ferns belong here.

Common Fern Family, POLYPODIACEAE

Figure 262

Fig. 262. a, *Asplenium filixfoemina* (L.), Lady Fern. Common throughout much of our country; b, *Adiantum pedatum* L., Maiden-hair Fern. A well-known beautiful deep woods fern; c, *Pellaea atropurpurea* (L.), Purple-stemmed Cliff-brake. Grows in crevices on limestone bluffs.

Ferns are highly favored as ornamentals yet some species become serious weed pests. There are in all close to 5000 species known in this family.

ORDER LYCOPODIALES

9a Aquatic or mud plants; leaves hollow, cylindrical; stems short.

Quillwort Family, ISOETACEAE

Figure 263

Fig. 263. *Isoetes engelmanii* A. Br., Engelman's Quillwort.

The tapering quill-like leaves, 20 to 100, grow attached to a short fleshy stem and stand 8 to 15 inches high. It roots in mud with part of the plant usually submerged. Spores are of two kinds, as shown and are borne in a cavity in the leaf base.

9b Terrestrial plants, leaves moss-like; stems elongate, creeping.....10

10a Spores all of one kind and size; usually larger and coarser plants
than 10b.   Club Moss Family, LYCOPODIACEAE

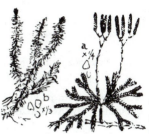

Figure 264

Fig. 264. a, *Lycopodium complanatum* L.,
Ground-pine. Erect stems arising 5-8
inches high from trailing stems several
feet in length; b, *Lycopodium lucidulum*
Michx., Shining Club-moss, 6 to 10 inches
high.

A good number of the something over
100 species constituting this family are
tropical *(epiphytes)*; while others are terrestrial. Those in our region grow in deep
damp woods. Club mosses are used for
Christmas decorations and are often put
into ornamental wreaths.

10b Spores of two kinds and sizes, small microspores developing male
gametophytes and quite large megaspores.
Little Club Moss Family, SELAGINELLACEAE

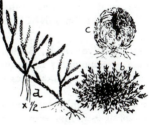

Figure 265

Fig. 265. a, *Selaginella rupestris* (L.),
Rock Selaginella. Rather common on
dry rocks, over 2 or 3 inches high;
Selaginella lepidophylla, Resurec-
tion plant; b, during rainy season; c,
during dry weather. Rather common
in the Southwest.

Some very beautiful members of this
family are cultivated as "table ferns".

THE SPERMATOPHYTA (SEED-BEARING PLANTS)

1a Ovules and seeds borne on the surface of a bract or scale; no
stigma. Figs. 266 to 269. Class I GYMNOSPERMAE. 2

1b Ovules and seeds in a closed cavity (ovary) formed by the union
of one or more modified leaves, the tips of which have developed
into one or more stigmas for the reception of pollen. (Includes all of
our common flowering plants). Class II ANGIOSPERMAE. 4

THE GYMNOSPERMS

2a (a, b, c) Leaves pinnately compound; evergreen; trunk usually short and unbranched.   Cycas Family, CYCADACEAE

Figure 266

Fig. 266. *Cycas revoluta* Thunb., Sago Palm; a, an old plant (female); b, a fruiting leaf with ovules.

This is not a true palm, of course, as the palms are Monocotyledons (see Fig. 270). This cycad reaches a height of 6 to 10 ft. and its leaves a length of 2 to 7 ft. Cycads develop slowly. Fossil remains show that they were once very abundant. Younger plants have their leaves close to the ground.

2b Leaves fan-shaped, deciduous on large spreading trees.   Ginkgo Family, GINKGOACEAE

Figure 267

Fig. 267. *Ginkgo biloba* L., Maidenhair Tree; a, twig with dwarf branch, leaves and male flowers; b, fruit; c, female flowers and young fruit.

There is but this one species known for this family but it has found an important place as an ornamental tree.

2c Leaves simple, scale-like, awl-shaped or needle-like; usually evergreen. Shrubs and trees. 3

3a Seeds borne under the scales of dry cones or of somewhat fleshy berry-like cones.   Pine Family, PINACEAE

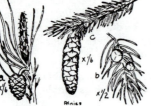

Figure 268

Fig. 268. a, *Pinus sylvestris* L., Scotch Pine; b, *Juniperus communis* L., Common Juniper; c, *Picea glauca* Voss, White Spruce.

This is one of the highly important plant families as it contains many of our most valuable lumber trees.

3b Seeds in a fleshy cup. Flowers single or in pairs. Trees or shrubs.   Yew Family, TAXACEAE

Figure 269

Fig. 269. *Taxus bacata* L., English Yew. a, branch with fruit; b, fruit; c, young cone.

A tree reaching a height of 60 ft.; often cultivated. The Oregon Yew is a large forest tree while the American Yew (*Taxus canadensis*) is only a straggling shrub seldom attaining a height of 5 ft.

THE ANGIOSPERMS

4a Leaves usually parallel-veined (a); flowering parts usually in 3's (b); stems hollow or with bundles scattered throughout (c); seeds usually with but one cotyledon. Fig. 270.

THE MONOCOTYLEDONS. 5

Figure 270

4b Leaves usually net-veined (a); flowering parts usually in 5's (sometimes 4's) (b); stems with bundles arranged in a ring surrounding the pith, or woody with outer bark; seeds usually with two cotyledons. Fig. 271.

THE DICOTYLEDONS page 100

Figure 271

THE MONOCOTYLEDONS

5a Plants with the palm-type foliage, with large pinnate or palmate compound leaves. Palm Family, **PALMACEAE**

Figure 272

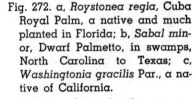

Fig. 272. a, *Roystonea regia*, Cuba Royal Palm, a native and much planted in Florida; b, *Sabal minor*, Dwarf Palmetto, in swamps, North Carolina to Texas; c, *Washingtonia gracilis* Par., a native of California.

This is a large family comprising almost 200 genera and more than 1500 species of mostly tropical evergreen plants. The majority are trees, some becoming quite large, but some members of the family are shrubs and others great climbers. The palms seen in greenhouses are usually juvenile plants and too immature to be easily determined to genus or species. The family yields many products of high economic importance. About 20 species of palms are native of the U. S.

5b Plants not as in 5a. Figs. 273 to 302. 6

6a Small circular oval or flash-shaped plants floating free on water. No true leaves. Duckweed Family, **LEMNACEAE**

Figure 273

Fig. 273. a, *Spirodela polyrhiza* (L.) Greater Duckweed; b, *Lemna trisulca* L., Ivy-leaved Duckweed.

These tiny plants produce flowers and seeds but one must look very close to see these structures. The winter is passed by sinking to the bottom of the pond.

91

6b Plants not as in 6a. Figs. 274 to 302. 7

7a Flowers minute, surrounded by chaffy bracts (glumes); without a 3-parted perianth; flowers grouped in spikes or spikelets. Fig. 274. (The Grasses and Sedges). 8

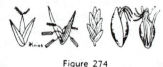

Figure 274

7b Flowers not in chaffy bracts or scales. Figs. 277 to 302. 9

8a Leaves two-ranked (in two rows on the stem), the edges of their sheathes not united; stems cylindrical or flattened and almost always hollow; anthers attached by their middle; fruit a grain.
Grass Family, GRAMINEAE

Figure 275

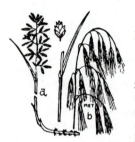

Fig. 275. a, *Poa compressa* L., English Blue-grass; b, *Bromus tectorum* L., Downy Brome-grass.

This is one of the largest and also one of the most important of all plant families. Removal of this family would cut off a very large percentage of the food crops of the world. Common examples: Corn, Wheat, Rice, Oats, Sugar Cane, Bamboo, etc.

8b Leaves three-ranked (in 3 rows on the stem), the edges of their sheaths united; stems almost always triangular and solid; anthers attached at one end; fruit an achene.
Sedge Family, CYPERACEAE

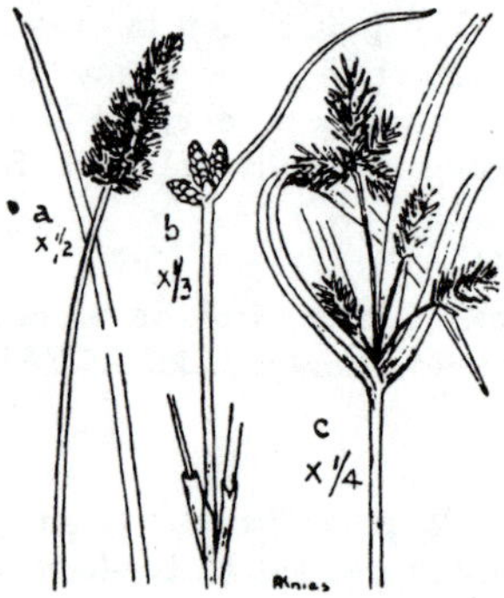

Figure 276

Fig. 276. a, *Carex aggregata* Mack.; Glomerate Sedge; b, *Scirpus debilis* Pursh., Club Rush; c, *Cyperus erythrorhizos* Muhl., Red-rooted Cyperus.

The sedges resemble the grasses in a superficial way. While they constitute a large family they are of comparatively small economic importance though highly interesting.

9a Flowers with only a rudimentary perianth, sometimes bristles or scales, or none. Figs. 277 to 280. 10

9b Flowers with sepals and petals (sometimes the two are quite similar). Figs. 281 to 302. 13

10a Flowers in a fleshy spike (spadix) (a^1) arising from an enlarged bract (spathe) (a^2).   **Arum Family, ARACEAE**

Fig. 277. a, *Arisaema dracontium* (L.), Green Dragon; b, *Acorus calamus* L., Sweet Flag or Calamus.

The leaf-like part projecting beyond the flowering spike is the spathe. The drug calamus comes from the heavy root stalks of this plant. Common Examples: Calla Lilies, Elephant's Ear.

Figure 277

10b Flowers not as in 10a. Figs. 278 to 280. 11

11a Submerged aquatic plants, leaves often floating.
                              Pondweed Family, NAJADACEAE

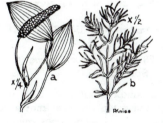

Fig. 278. a, *Potamogeton heterophyllus* Schreb., Various-leaved Pondweed; b,

Najas flexilis (Willd.), Slender Najas.

These are exceedingly important from the standpoint of maintaining a balanced condition of life in our water courses. They supply great quantities both of oxygen and of food for the many plant-feeding aquatic animals.

Figure 278

11b Not normally submerged although usually growing in marshy places. Figs. 279 and 280. 12

12a Flowers in elongate terminal spikes, fruit surrounded by bristles.
                      Cat-tail Family TYPHACEAE

Fig. 279. *Typha latifolia* L., Broad-leaved Cat-tail.

Cat-tails are found in marshes almost everywhere. Immense quantities of pollen is produced in the upper cylinder, while the lower part of one stalk may contain more than a million seeds.

Figure 279

12b Flowers in spherical lateral spikes.

      Bur-reed Family, SPARGANIACEAE

Fig. 280. *Sparganium eurycarpum* Engelm., Broad-fruited Bur-reed.

This family contains but the one genus and only twenty some species. Most of the species are from one to 3 ft. in height but the species pictured may attain a height of 8 ft.

Figure 280

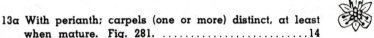

13a With perianth; carpels (one or more) distinct, at least when mature. Fig. 281. 14

Figure 281

13b Carpels, one or more, united into one compound ovary. Fig. 282. 15

Figure 282

14a Petals and sepals similar, anthers long and narrow; carpels attached to each other until ripe..

      Arrow-grass Family, SCHEUCHZERIACEAE

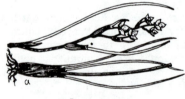

Fig. 283. a, *Triglochin maritima* L., Spike-grass.

This small family, — there are only some ten species known to science,—is of small consequence except that it must be accounted for somewhere.

Figure 283

14b Petals and sepals unlike, petals usually white; anthers short and thick; carpels separate.

      Water-plantain Family, ALISMACEAE

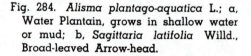

Fig. 284. *Alisma plantago-aquatica* L.; a, Water Plantain, grows in shallow water or mud; b, *Sagittaria latifolia* Willd., Broad-leaved Arrow-head.

Many members of this family have leaves shaped like arrowheads, hence the name.

Figure 284

94

15a Ovary superior (hypogynous;-above the calyx). Fig. 285. 16

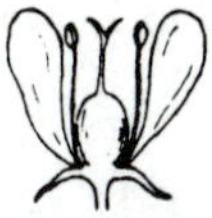

Figure 285

15b Ovary wholly or partly inferior (epigynous;-below the calyx). Fig. 286. 21

Figure 286

16a Some or all of the members of the perianth scaly, chaffy bracts. Figs. 287 to 289. 17

16b Not as in 16a. All members of the perianth leaf-like or flower-like in texture. Figs. 290 to 302. 19

17a Perianth of six chaffy, scale-like similar parts; leaves grass-like; flowers small, perfect.   The Rush Family, JUNCACEAE

Fig. 287. a, *Juncus tenuis* Willd., Path Rush, Common throughout North America.

Watched by the Indians in following old pathways; b, *Juncus brachycarpus* Engelm., Short-fruited, Rush.

Figure 287

17b Some of the perianth lobes chaffy, others not; herbs with erect stems. Figs. 288 and 289. 18

18a Flowers perfect (bisexual), yellow; in terminal scaly heads or spikes.   Yellow-eyed Grass Family, XYRIDACEAE

Fig. 288. *Xyris carolinana* Walt., Carolina Yellow-eyed Grass.

This species inhabits swamps and bogs and is largely confined to the Atlantic coast from Massachusetts south. Bogs form the usual habitat of this small family although an occasional species lives on dry land.

Figure 288

18b Flowers imperfect (unisexual) in terminal scaly heads.
Pipewort Family, ERIOCAULACEAE

Fig. 289. a, *Eriocaulon articulatum* (Huds.), Seven-angled Pipewort.

The members of this family are usually perennials and aquatic. They grow in bogs or similar damp regions.

Figure 289

19a (a, b, c) Terrestrial herbs. Figs. 292 to 302. 20

19b Aquatic herbs, stamens 3 or 6 partly attached to perianth; perianth of 6 parts all much alike, tubular.
Pickerel-weed Family, PONTEDERIACEAE

Fig. 290. a, *Pontederia cordata* L., Pickerel-weed. G r o w s along streams and ponds; the flowers are bright blue; b, *Heteranthera dubia* (Jacq.), Water Star-grass.

Grows in water and occasionally along shore. Flowers are light yellow.

Figure 290

19c Plants usually epiphytic, often moss-like. Leaves and stems usually scurfy or mealy.
Pineapple Family, BROMELIACEAE

Fig. 291. a, *Tillandsia* sp., Air Pine; b, *Dendropogon usneoides* (L.) Spanish Moss; c, *Ananas comosus* Merr., Pineapple.

Most of the members of this large family are confined to the tropics. The pineapple is one of the few members not ephiphytic. Florida (or Spanish) "moss" (it is not a moss of course) drapes a majority of the trees throughout the south. It is used for mattress and automobile upholstery.

Figure 291

20a Stamens usually 6; perianth of similar, mostly colored divisions (often more or less complete united); or 3 green sepals and 3 colored, withering persistent petals; plants often growing from bulbs; seed with horny endosperm. Lily Family LILIACEAE

Fig. 292. a, *Lilium canadense* L., Nodding Lily; b, *Trillium grandiflorum* .Sals., Large-flowered Wake Robin; c, *Asparagus officinalis* L., Asparagus.

Common examples: Onion, Garlic, and many ornamental plants.

Figure 292

20b Periants of 3 persistent, usually green sepals and 3 ephemeral colored petals. Stamens 6, free; similar or dissimilar.

Spiderwort Family, COMMELINACEAE

Fig. 293. a, *Commelina communis* L., Asiatic Day-flower. Flowers bright blue. Widely distributed. Introduced from Asia; b, *Tradescantia reflexa*, Ruf. Spiderwort.

The Spiderworts, very common both wild and in cultivation are showy plants. The Wandering Jew, a frequent ornamental belongs here.

Figure 293

21a Flowers dioecious (having only stamens or carpels (imperfect) but both kinds on the same plant). Figs. 294 and 295. 22

21b Flowers perfect (both stamens and carpels in the same flower); ovules and seeds usually numerous. 23

22a Aquatic herbs. Stamens 3-12 sometimes united together; ovules and seeds several or numerous.

Frog's Bit Family, HYDROCHARITACEAE

Fig. 294. a, *Elodea canadensis* Michx., Water-weed; b, *Vallisneria spiralis* L., Eel Grass. The above two plants are much used in aquaria as aerators and are sold for this purpose; c, *Limnobium spongia* (Bosc.) Frog's Bit. Flowers white.

Figure 294

22b Terrestrial twining plants; ovules and seeds but one or two in each of 3 cavities of ovary. Yam Family, DIOSCOREACEAE

Figure 295

Fig. 295. *Dioscorea villosa* L., Wild Yam-root.

There is but this one native species known for this family in the northern U. S. Some 200 species are found in warmer areas. Many of them grow large edible roots or above-ground tubers which make an important food contribution. (Some sweet potatoes are known as "Yams" but they belong to the family Convolvulaceae, Fig. 455).

23a (a, b, c) Fertile stamens 3 or less. Figs. 298 to 302. . . . .**24**

23b Fertile stamens (maturing pollen) 6.
Amaryllis Family, AMARYLLIDACEAE

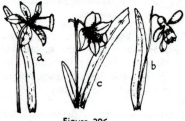

Figure 296

Fig. 296. a, *Narcissus pseudonarcissus* L. Daffodil; b, *Galanthus elwesii* Hook f., Giant Snowdrop; c(*Hippeastrum vittatum* (Herb.), Common Amaryllis.

At first glance the members of this family seem to be lilies but they differ in having the ovary below the flower instead of up in the flower as in the lilies.

23c Fertile stamens 5, sterile stamen (staminode) 1, not petal-like; plants with very large leaves. Banana Family, MUSACEAE

Figure 297

Fig. 297. *Musa paradisiaca sapientum* L., Banana.

The banana stalk attains a height of 15 to 30 ft. and bears one bunch of fruit when about a year old and then dies but the suckers grow into new fruit-bearing plants. Bananas may make the amazing yield of 100 tons per acre.

Other common example: Manila Hemp.

24a Fertile stamens only one or two. Figs. 300 to 302. 26

24b Fertile stamens 3. Figs. 298 and 299. 25

25a Stamens opposite the outer perianth segments; anthers facing outward and opening lengthwise.   Iris Family, IRIDACEAE

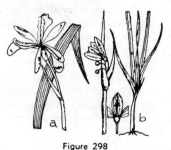

Figure 298

Fig. 298. a, *Iris versicolor* L., Larger Blue Flag; b, *Sisyrinchium angustifolium* Mill., Blue-eyed Grass.

Over 170 species of Iris are known and innumerable varieties.

25b Stamens opposite the inner perianth segments; anthers usually versatile. Perianth woolly. Bloodwort Family, HAEMODORACEAE

Figure 299

Fig. 299. *Lachanthes tinctoria* (Walt.), Redroot.

In swamps along eastern part of our country; flowers yellow.

26a Stamens separate from the carpels. Figs. 301 and 302. 27

26b Stamens 1 or 2 united with the carpels. Flowers bilaterally symmetrical or very irregular; seeds fine and numerous.
          Orchid Family, ORCHIDACEAE

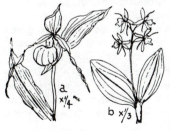

Figure 300

Fig. 300. a, *Cypripedium hirsutum* Mill., Large Yellow Lady's Slipper; b, *Liparis loeselii* Rich., Fen Orchis.

This family likely has more species than any other family of plants. More than 15,000 species are now known. The majority are tropical and many are epiphytic.

27a Ovules many in each cell; fruit a capsule with hard spherical seeds. Fertile stamens one with one or more sterile petal-like stamens.

Canna Family, CANNACEAE

Fig. 301. a, *Canna indica* L., Indian Shot, flowers small, leaves very large; b, *Canna generalis* Bailey, Common Flowering Canna.

This family with its one genus of some 40 species has very highly specialized flowers. They are mostly tropical but are raised extensively for their showy foliage and flowers.

Figure 301

27b Ovules but one in each cell; fruit a capsule or berry-like, 1 to 3 celled. Fertile stamens 1, petal-like sterile stamens often 5.

Arrowroot Family, MARANTACEAE

Fig. 302. *Maranta leuconeura* Morr.

An ornamental from Brazil, prized for its showy leaves.

Common example: Arrowroot (*Maranta arundinacea* L.) a source of tapioca.

Figure 302

THE DICOTYLEDONS

1a Corolla (petals) none; calyx (sepals) present or absent, sometimes colored and resembling a corolla. Figs. 303 to 331. 2

1b Both calyx and corolla present, at least in the staminate or perfect flowers. Figs. 332 to 472. 25

2a Trees with branches jointed resembling equisetum; only minute scales for leaves.   Casuarina Family, CASUARINACEAE

Fig. 303. *Casuaria equisetifolia* L., Beefwood, Australian Pine.

a, branches with long hair-like green stems; b, and c, sections of young stems showing scale-like leaves.

These trees are native of the Australian region but are extensively planted in our southern states for ornament, windbreaks, etc. Rapid growers; wood hard.

Figure 303

2b Plants with normal stems and leaves. Figs. 304 to 331. 3
3a Calyx present; sepals distinct or united, green or colored.
Figs. 312 to 331. 9
3b Without a perianth (though a cup, gland or minute border may be
present in the place of a calyx). Figs. 304 to 311. 4
4a (a, b, c, d) Herbs fairly large and sturdy (in marshes).
Lizard's Tail Family, SAURURACEAE

Fig. 304. *Saururus cernus* L., Lizard's Tail.
This family has broad alternate leaves and 3-4 carpels in an ovary with 1 to 2 seeds in each carpel. The Pepper Family (Piperaceae) closely related tropical and greenhouse plants are distinguished by having but one seed in each ovary. It includes woody as well as herbaceous plants.

Figure 304

4b Small frail herbs growing in mud or water. Leaves entire, even
though submerged.   Water Starwort Family, CALLITRICHACEAE

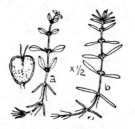

Fig. 305. *Callitriche palustris* L., Water Fennel; a, terrestrial form; b, when submerged.

This family has but one genus. Some grow on wet soil, others in water. There are about 20 species widely distributed.

Figure 305

4c Submerged aquatic plants with finely divided leaves and minute
flowers. Figs. 306 and 307. 5
4d Trees or shrubs; staminate and sometimes the carpellate flowers in
aments (slender spike of flowers). Figs. 308 to 311. 6
5a Plants 1 to several feet in length with finely cut whorled leaves;
stamens many, sessile. An involucre (a) cut into many parts may
be mistaken for a calyx.   Hornwort Family, CERATOPHYLLACEAE

Fig. 306. *Ceratophyllum demersum* L., Hornwort.
It is much used in aquaria for aeration. It is widely scattered in lakes and slow moving streams. There is but this one genus for the family and only a few species.

Figure 306

5b Plants small, resembling moss or liverworts; stamens (but 2 in our genus) with filaments.   **River Weed Family, PODOSTEMACEAE**

Fig. 307. *Podostemum ceratophyllum* Michx., River Weed.

The flower arises from a cup-like spathe with uncut edge. This olive green plant grows in streams where it is attached to the rocks.

Figure 307

6a Large trees with broad palmate leaves, the base of the petiole surrounding the bud.   **Plane-tree Family, PLATANACEAE**

Fig. 308. *Platanus occidentalis* L., Sycamore.

This family has but the one genus. The bark peels in large pieces from the upper limbs leaving the branches a pale greenish-white. Seeds are borne in round hanging spheres. "a" is the leaf of the London Plane sometimes planted as an ornamental.

Figure 308

6b Trees and shrubs, but with axillary bud not covered by leaf base. Figs. 309 to 311. ....**7**

7a Fruit many-seeded. Seeds when ripe carried by wind by tufts of hairs at one end.   **Willow Family, SALICACEAE**

Fig. 309. a, *Populus tremuloides* Michx., Quaking Aspen; b, *Salix interior* Rowl., Sandbar Willow.

This is an important family of quick-growing soft wood trees. Much used for firewood, cheap lumber, and training small children.

Figure 309

7b Seeds not tufted for wind dispersal. Figs. 310 and 311. ....**8**

8a Fruit one-seeded; seeds without tufts of hairs, leaves resin-dotted. Carpellate flowers solitary with but one carpel and one ovule.
Bayberry Family, MYRICACEAE

Figure 310

Fig. 310. a, *Myrica gale* L., Sweet Gale.

A widely distributed shrub growing in swamps and along watercourses.

Three genera and some 35 widely distributed species have been named.

8b Carpellate flowers with compound ovary of two carpels. (Several united in a pendulous ball in the Sweetgum). Fruit a woody capsule.
Witch Hazel Family, HAMAMELIDACEAE

Figure 311

Fig. 311. a, *Liquidambar styraciflua* L., Sweetgum; b, *Hamamelis virginiana* L., Witch Hazel.

It will be noted that the flowers of this plant have both sepals and petals. They appear in late fall; the seeds mature a year later.

9a Flowers, at least the staminate, in aments. Figs. 312 to 315. 10

9b Staminate flowers not in aments, but in clusters of different types, or rarely solitary. Figs. 316 to 331. 14

10a Plants parasitic on trees; fruit a berry.
Mistletoe. Family, LORANTHACEAE

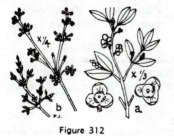

Figure 312

Fig. 312. a, *Phoradendron flavescens* (Pursh), American Mistletoe.

Grows on several broad-leaved trees of East and South; b, *Razoumofskya americana* (Nutt.), American Dwarf Mistletoe. Parasitic on western pines. The traditions relating to Mistletoe make is very helpful for the socially shy.

10b Plants not parasitic. Figs. 313 to 315. 11

11a Both staminate and pistilate flowers in aments; neither a cup nor bur with the fruit.　　　　　　　　　Birch Family, BETULACEAE

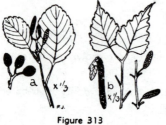

Figure 313

Fig. 313. a, *Alnus tenuifolia* Nutt., Thin-leaved Alder. Along watercourses in West; b, *Betula populifolia* Marsh., Gray Birch.

It is worth driving long distances just to see the great birch forests of our northern states and Canada.

11b Only the staminate flowers in aments. Figs. 314 and 315. 12

12a Leaves pinnately compound; fruit a nut with husk.
　　　　　　　　　　　　　　　Walnut Family, JUGLANDACEAE

Figure 314

Fig. 314. a, *Juglans nigra* L., Black Walnut; b, *Carya ovata* Koch., Shagbark Hickory.

This is a highly important family as it produces some of our most valuable woods and its nuts have good market value.

12b Leaves simple. Figs. 315 and 317. 13

13a Fruit a nut with bur or cup.　　　　　Beech Family, FAGACEAE

Figure 315

Fig. 315. a, *Quercus alba* L., White Oak; b, *Fagus grandifolia* Ehrh., American Beech.

The oaks produce some of our strongest timbers, finest finishing lumber and most beautiful trees. It is a grand genus.

13b Fruit a collection of tiny drupes uniting at maturity into a single somewhat-fleshy fruit (Mulberries, etc.). (See Fig. 317a).
　　　　　　　　　The Nettle Family (in part), URTICACEAE

14a (a, b, c) Ovary superior, flowers largely monoecious.
Figs. 317 and 318. 15

14b Ovary superior, flowers perfect.
Fig. 316a. 17

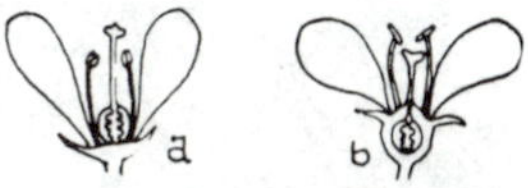

14c Ovary inferior. Fig. 316b. 16

Figure 316

15a Ovary 1-celled; fruit one-seeded. The Nettle Family, URTICACEAE

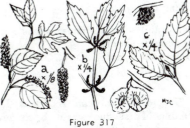

Figure 317

Fig. 317. a, *Morus rubra* L., Red Mulberry; b, *Pilea pumila* (L.), Clearweed; c, *Ulmus fulva* Michx., Slippery Elm.

The American Elm is perhaps our most popular shade tree. Elms also furnish valuable lumber though not so desirable as oak.

15b Ovary usually 3-celled with one to 2 ovules in each cell; staminate and pistilate flowers on the same spike.

Box Family, BUXACEAE

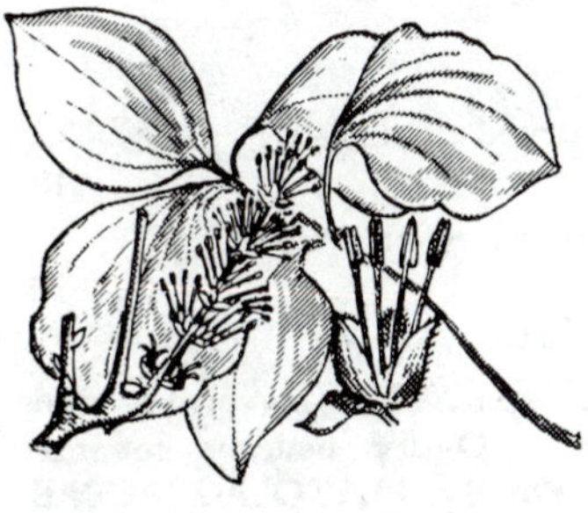

Figure 318

Fig. 318. *Pachysandra procumbens* Michx., Alleghany Mountain Spurge.

Common throughout our eastern mountains. The Common Box, an old world shrub or small tree, is a much prized ornamental which belongs to this family.

16a Ovary one-celled.
Sandalwood Family, SANTALACEAE

Fig. 319. a, *Comandra pallida* A. Dc., Pale Comandra. Erect herb about one foot high partially parasitic on roots of other plants, flowers purplish; b, *Pyrularia pubera* Michx., Oil-nut, a shrub 3 to 15 ft. tall.

Sandalwood trees and shrubs, native of India and Australia, are sources of oil used in perfume.

Figure 319

16b Ovary several-celled (usually 6); flowers perfect.
Birthwort Family, **ARISTOLOCHIACEAE**

Figure 320

Fig. 320. a, *Aristolochia tomentosa* Sims., Woolly Pipe-vine. The plant world has about 200 species of this genus. They are climbers for the most part and mostly woody. The flowers take peculiar shapes and in some species are large and highly colored. b, *Asarum canadense* L., Wild Ginger. A beautiful plant with velvety leaves and purplish-brown flowers. The stout rhizome has a strong odor of ginger.

17a Fruit an achene (dry one-seeded fruit); leaves simple; stems surrounded above each node by a stipular sheath.
Buckwheat Family, **POLYGONACEAE**

Figure 321

Fig. 321. a, *Polygonum scandens* L., Climbing False Buckwheat; b, *Rumex altissimus* Wood, Tall Dock.

Common examples: Spinach-Dock, Rhubarb, Buckwheat, Prince's Feather, and many bad weeds.

19a Ovary severel celled. Fruit (in our species) a several-seeded berry with staining juice. Large herb. Ovary usually several-celled.
Pokeweed Family, **PHYTOLACCACEAE**

Figure 322

Fig. 322. *Phytolacca decandra* L., Common Poke.

The Indians used Poke berries for paint. They make a beautiful shade of magenta. The roots are said to be poisonous.

19b Ovary one-celled, one-seeded fruit a berry-like drupe; shrubs or trees. Figs. 323 to 325. 20

20a Leaves with translucent dots. Sepals 4-6 stamens 9 to 12. Aromatic trees or shrubs.   **Laurel Family, LAURACEAE**

Figure 323

Fig. 323. a, *Sassafras albidum* Nees, Sassafras.

The outer covering of the roots of the Sassafras tree furnishes the bark from which spring remedies have long been made. Sometimes called the "Mitten-tree" because of the shape of some of its leaves.

20b Leaves with silvery, scurfy scales. Shrubs or small trees. Sepals 4; stamens 4 to 8 (petals, in some exotic species).

Oleaster Family, ELAEAGNACEAE

Fig. 324. *Elaeagnus argentea* Pursh., Silverberry.
Leaves, fruit and outside of flowers silvery. Flowers yellow within. The Russian-olive, a related species, makes a valuable windbreak in northern regions.

Figure 324

20c Shrubs with very tough bark. Calyx with 4 or 5 lobes or none. (Petals, in some exotic species). Mezereum Family, THYMELACEAE

Fig. 325. *Dirca palustris* L., Leather-wood.

The tough yellowish-green stems were used by the Indians as cords for tying.

The family has 40 genera and over 400 species but they are mostly confined to the Eastern Hemisphere.

Figure 325

21a (a, b, c) Fruit a capsule, more than one-celled. (If only one-celled, see Portulacaceae, Figs. 349 and 362, and Caryophyllaceae, Figs. 373 and 396, which have a few species without petals).

Carpet-weed Family, AIZOACEAE

Figure 326

Fig. 326. *Mollugo verticillata* L., Carpet-weed.

It grows very flat to the ground and seems to do best where there is not much competition. New Zealand Spinach belongs to this family.

21b Fruit a 1-celled capsule, a nut, or a drupe; with calyx, a long cylindrical tube usually swollen at base. Native in the southern hemisphere and tropics.   **Family PROTEACEAE**

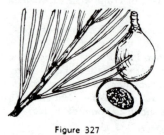

Fig. 327. *Hakea* sp.

A fairly large family. Several species such as the Silver Tree, Silk Oak, Banksia and Queensland Nut are planted as ornamentals.

Figure 327

21c Fruit not a more than 1-celled capsule. Figs. 328 to 331. 22

22a Calyx colored like a corolla; flowers in clusters surrounded by an involucre of distinct or united bracts.

Four-o'clock Family, NYCTAGINACEAE

Fig. 328. a, *Oxybaphus nyctagineus* Sweet, Wild Four-o'clock, a rather common roadside weed with pink flowers; b, *Buginvillaea glabra* Choisy.

Some important ornamentals belong to this family.

Figure 328

22b Calyx not corolla-like. Figs. 329 to 331. 23

23a Leaves with scarious stipules (in one genus stipules absent, but stamens arise from margin of a hypanthium).

Knotwort Family, ILLECEBRACEAE

Fig. 329. a, *Anychia canadensis* (L), Forked Chickweed; b, *Paronychia dichotoma* (L.), Forking Whitlow-wort.

These plants have tiny white or greenish flowers.

More than 100 species of this family are known, many of them widely distributed.

Figure 329

23b Leaves without stipules. Figs. 330 and 331. 24

24a Bracts dry, harsh and not green, perianth parts harsh, sharp-pointed.   **Amaranth Family, AMARANTHACEAE**

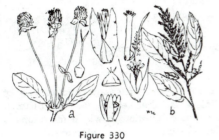

Figure 330

Fig. 330. a, *Amaranthus retro-flexus* L., Red Root; b, *Gomphrena globosa* L., Globe Amaranth, a popular cultivated ornamental; flowers red, lavender, or white. They are dried and used in permanent bouquets.

24b Flowering parts and bracts usually soft and green; plants in many cases white-mealy.   **Goosefoot Family, CHENOPODIACEAE**

Figure 331

Fig. 331. a, *Salsola pestifer* Nelson, Russian-thistle, a widely distributed serious pest from Europe and Asia. b, *Chenopodium album* L., Lamb's Quarters.

Common examples: Sugar and Garden Beet (same species), Common Spinach, and Summer Cypress.

25a Petals separated from each other. Fig. 332. 26

Figure 332

25b Petals at least partly united. Fig. 333. 105
In many members of the Evening Primrose family the petals arise from the top of a long calyx tube and might seem to be "united". Take 25a or go to 96 and 104.

Figure 333

26a Ovary superior (sepals arising below the ovary). Fig. 316a. 27

        (The peas and their many relatives with a two-halved, one-celled ovary and usually irregular flowers, might seem to go here. If so turn to 36 or 115.)

26b Ovary inferior or at least partly so, (sepals arising from sides or top of ovary). Fig. 316b. 92

27a Carpels distinct, one to many. Fig. 334a. 28

27b Carpels more than 1, united into a compound ovary. Fig. 334b. 39

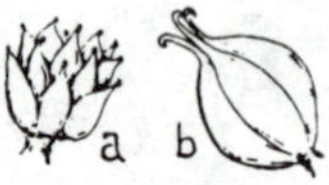

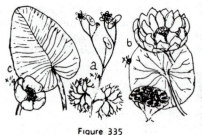

Figure 334

28a Stamens usually more numerous than the sepals and arising at the base of the ovary or below it; the sepals usually distinct. (There are many double forms in cultivation in which the stamens have changed into petals leaving but few or no stamens.) Figs. 335 to 342. 29

28b Stamens usually arising around or above the ovary, sepals usually on the edge of a cup-like receptacle. Figs. 343 to 346. 36

29a Aquatic herbs; carpels 3 or more, petals large; usually large floating peltate leaves.　　　Water Lily Family, NYMPHAEACEAE

Fig. 335. a, *Cabomba caroliniana* A. Gray, Cabomba; b, *Nelumbo lutea* (Willd.), American Lotus; c, *Nymphaea advena* Soland, Cow Lily.

This and the genus *Castalia* have the carpels united in a compound pistil. Many exotic water lilies with beautifully tinted flowers are available for lily pools.

Figure 335

29b Growing on land or in marshes (a few species of Ranunculaceae submerged), stamens many; sepals distinct. Figs. 336 to 342. 30

30a (a, b, c) Receptacle hollow, containing many carpels and their achenes, roughly appearing as a compound ovary. Aromatic shrubs, with opposite entire leaves.

Calycanthus Family, CALYCANTHACEAE

Fig. 336. a, *Calycanthus floridus* L., Hairy Strawberry Shrub.

So-called because of the odor of strawberries given off by the dark purplish-red flowers. A native of southeastern United States but much planted as an ornamental.

Figure 336

110

30b Receptacle solid; sepals and the many stamens on a ring (hypanthium) surrounding the carpels; leaves usually with stipules. See Figs. 345, 409 and 411½. Rose Family, ROSACEAE

30c Receptacle not hollow. Leaves almost always alternate. Figs. 337 to 342. 31

31a Climbing vines with simple alternate leaves and small dioecious flowers. Moonseed Family, MENISPERMACEAE

Fig. 337. *Menispermum canadense* L., Canada Moonseed.

Flowers white, drupes bluish-black resembling a bunch of small grapes. Most of the members of the family are tropical.

Figure 337

31b Not vines with simple leaves, or if so, then the flowers are perfect. Figs. 338 to 342. 32

32a Trees and shrubs with large entire (rarely lobed) alternate leaves; sepals 3. Figs. 338 and 339. ....3?

32b Not as in 32a. Figs. 340 to 342. 34

33a Sepals (3) and petals (6) meeting at their edges in the bud (valvate) flowers not showy; fruit large, fleshy elongated berries. Custard Apple Family, ANONACEAE

Fig. 338. *Asimina triloba* Dunal, Pawpaw.

The inter-dependence of plants and animals is well illustrated here. The Zebra butterfly, one of our most attractive swallowtails, feeds in its larval state only on the Pawpaw. To look for it in regions where Pawpaw does not grow is a waste of time.

Figure 338

33b Sepals (3) and petals (6 to 12) overlapping in the bud (imbricate); flowers fragrant and showy; fruit cone-like. Magnolia Family, MAGNOLIACEAE

Fig. 339. a, *Liriodendron tulipifera* L., Tuliptree; b, *Magnolia virginiana* L., Sweetbay.

These plants naturally belong to the southern states, but the Tuliptree and a few species of magnolias can be raised in the north if given a bit of extra care.

Figure 339

34a Carpels usually more than 1; sepals 3-15 overlapping in the bud; petals occasionally 0 but usually 3 to many; fruit an achene, follicle or berry; anthers not opening by valves.

Crowfoot Family, RANUNCULACEAE

Fig. 340. a, *Delphinium tricorne* Michx., Dwarf Larkspur; b, *Ranunculus acris* L., Meadow Buttercup. Often cultivated, sometimes very double. c, *Actaea rubra* (Ait.), Red Baneberry.

Other common examples: Peony, Clematis, Anemone, and Meadow Rue. All of which are favorite garden ornamentals.

Figure 340

34b Carpels only 1. Figs. 341 and 342. 35

35a Sepals more than 2; fruit a berry or capsule; anthers opening by valves. Barberry Family, BERBERIDACEAE

Fig. 341. a, *Berberis vulgaris* L., European Barberry. Outlawed throughout the wheat belt as it is the alternating host of Black Stem Rust of wheat. b, *Podophyllum peltatum* L., May Apple.

Often very abundant in low woods. The fruit is edible. When ripe it has a very characteristic flavor.

Figure 341

35b Sepals 2; climbing herbs with tubers at roots or on the vines. Basella Family, BASELLACEAE

Fig. 342. *Boussingaultia baseloides* H. B. K., Madeira-Vine.

A rapid-growing vine with white flowers and little tubers in axil of leaves. Native of tropical America, but running wild in southern U. S. Often planted as an ornamental vine.

Figure 342

112

36a Fruit a legume (pea-like pod); flowers usually sweet pea-shaped (a few are regular); leaves usually compound.

Pea Family, LEGUMINOSAE

Figure 343

Fig. 343. *Vicia micrantha* Nutt., Small-flowered Vetch; b, *Cassia chamaecrista* L., Partridge Pea; c, *Trifolium repens* L., White Clover.

This family furnishes some highly important foods for man and beast.

Common examples: Peas, Beans, Peanuts, Alfalfa, and many beautiful and valuable trees.

36b Not as in 36a. Figs. 344 to 346. ....**37**

37a Herbs, with perfectly symmetrical flowers (sepals, petals and carpels of same number and stamens of same or double number); leaves without stipules; fruit a follicle.

Orpine Family, CRASSULACEAE

Figure 344

Fig. 344. *Sedum purpureum* L., Live-for-ever.

This species introduced from Europe is common in cultivation and as an escape. A number of species of this family are often employed for rock gardens. They are highly drought resistant.

37b Not as in 37a. Figs. 345 and 346. ....**38**

38a (a, b, c) Regular flowers with many stamens (rarely few) and the 5 petals often notched; leaves with stipules; seeds with no endosperm. (Many cultivated forms are double as the result of stamens turning into petals. Such forms may have few or no stamens).

Rose Family, ROSACEAE

Figure 345

Fig. 345. a, *Geum canadense* Jacq., White Avens; b, *Prunus nigra* Ait., Canada Plum; c, *Rosa virginiana* Mill., Wild Rose.

A large and highly important family for beauty and use.

Common examples for the family: Blackberries, Plums, Cherries, Apples, Quinces, Strawberries, Roses, and Spirea.

38b Stamens seldom more than 10, usually less; seeds with endosperm; leaves often without stipules.

Saxifrage Family, SAXIFRAGACEAE

Fig. 346. a, *Heuchera hispida* Pursh., Rough Alum-root; b, *Ribes vulgare* Lam., Common Garden Currant.

Common examples: Deutzias, Syringas, Currants, and Gooseberries.

Figure 346

38c Shrubs and trees with simple alternate leaves and fruit a woody capsule. (Often without petals). See Fig. 311.

Witch-Hazel Family, HAMAMELIDACEAE

39a Stamens many (more than 10 and more than twice the sepals or calyx lobes). Figs. 347 to 357. 40

39b Stamens 10 or less. (Not more than twice the petals). Figs. 358 to 402. ....:.50

40a Ovary raised above corolla by a stalk; corolla with a fringed crown. Vines with tendrils, or erect herbs. Fruit a berry with many seeds. Passion Flower Family, PASSIFLORACEAE

Fig. 347. *Passiflora incarnata* L., Passion Flower.

Flowers are white or lavender with pink or purple crown. Native throughout the South. The fruit is edible.

Figure 347

40b Ovary sessile; flowers without a crown; no tendrils. Figs. 348 to 357. 41

41a Sepals 2; herbs. Figs. 348 to 350. 42

41b Sepals more than 2. Figs. 351 to 357. 44

42a Flowers perfect. Figs. 349 and 350. 43

42b Flowers with stamens or pistils only. Monoecious. Succulent tender herbs raised as house plants and in greenhouse.

Begonia Family, BEGONIACEAE

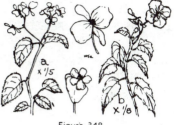

Figure 348

Fig. 348. a, *Begonia semperflorens* L. & O.; b, *Begonia tuberhybrida* Voss.

Many important house plants belong to this family, all falling in the one genus, *Begonia,* which is also the common name. The hybrid Tuberous-Rooted Begonias are among the most showy species.

43a Leaves thickened, succulent, entire (rarely 5 sepals).

Purslane Family, PORTULACACEAE

Figure 349

Fig. 349. a, *Portulaca grandiflora* Hook., Rose Moss. Flowers pink, red, yellow, white; often double; cultivated. Introduced from South America. b, *Talinum calysinum* Engelm., Large-flowered Talinum.

Many of the members of this family are xerophytes (belong naturally in regions of scanty rainfall).

43b Leaves often deeply cut or lobed, their tips often with spines. Juice milky or colored. Petals in pairs, 4-12. Flowers usually showy.

Poppy Family, PAPAVERACEAE

Figure 350

Fig. 350. a, *Papaver rhoeas* L., Corn Poppy. Flowers scarlet with dark center; b, *Sanguinaria canadensis,* L. Bloodroot. The sepals drop when the flower opens.

The Oriental poppies are perhaps the most showy of all the many species of poppies. Opium is made from the sap of one species.

44a Leaves with transparent or black dots. Stamens usually united in groups of 3's or 5's.

St. John's-wort Family, HYPERICACEAE

Fig. 351. *Hypericum prolificum* L., Shrubby St. John's-wort.

The flowers are usually yellow. The plants may be herbs or shrubs or sometimes small trees.

Figure 351

44b Leaves and stamens not as in 44a. Figs. 352 to 357. 45

45a Stamens united in one or a few large groups, sepals meeting at their edges. Figs. 352 and 353. 46

45b Stamens separate. Sepals overlapping in the bud. Figs. 354 to 357. 47

46a Stamens united in a central column surrounding the pistil (Fig. c).

Mallow Family, MALVACEAE

Fig. 352. a, *Hibiscus trionum* L., Flower-of-an-Hour; b, *Malva rotundifolia* L., Low Mallow.

Common examples: Cotton, Hibiscus, Hollyhock, and Indian Mallow (a most persistent weed).

Figure 352

46b Trees (in our region) with stamens in groups of five to ten each.

Linden Family, TILIACEAE

Fig. 353. *Tilia americana* L., American Basswood.

The wood is valued for combining whiteness, light weight, and fine grain. It is much used in cabinet work, and is a valuable honey plant. It is found throughout much of the eastern half of our country.

Figure 353

47a (a, b, c) Leaves hollow; pitcher, or trumpet shaped; herbs; in bogs.
Pitcher Plant Family, SARRACENIACEAE

Figure 354

Fig. 354. *Sarracenia purpurea* L., Pitcher Plant.

These plants are always interesting. *Sarracenia flava* L. commonly known as "Trumpets" grows abundantly in bogs of our southeastern states. The yellow and red leaves sometimes attain a height of three feet and are very conspicuous to man and doubtless also to the insects they trap.

47b Leaves solid, usually with stipules; sepals and the many stamens on a fleshy ring (hypanthium) surrounding the pistil. See Figs. 345, 409 and 411½. **Rose Family, ROSACEAE**

47c Not as in 47a. Figs. 355 to 357. .... **48**

48a Flowers not symmetrical. Figs. 356 and 357. .... **49**

Figure 355

48b Flowers symmetrical, leaves simple, almost always entire; sepals 3, or sometimes with an additional 2 small outer ones. **Rock-rose Family, CISTACEAE**

Fig. 355. a, *Helianthemum canadense* Michx., Rock-rose, flowers bright yellow; b, *Hudsonia ericoides* L., American Heath. Sandy soil in pine barrens suits it best.

Figure 356

49a Calyx persistent; open in bud; sepals and petals irregular, 4-8.
Mignonette Family, RESEDACEAE

Fig. 356. *Reseda lutea* L., Yellow Cut-leaved Mignonette.

The garden mignonette (*Reseda ordorata* L.) was an old time favorite largely because of its fragrance. Its leaves are entire or sometimes 3 lobed.

117

49b Sepals usually falling off, 4-8; petals 4; sap watery; leaves usually palmately compound.　　　　　Caper Family, CAPPARIDACEAE

Figure 357

Fig. 357. *Polanisia graveolens* Raf., Clammy-weed.

Common along sandy shores. The viscid pubescens of the leaves accounts for the common name.

The Spider Flowers, several species of which grow wild and are raised as ornamentals belong here.

51a Stamens united in 2 sets of 3; flowers irregular; sepals 2, scale-like.　　　　　Fumitory Family, FUMARIACEAE

Figure 358

Fig. 358. a, *Dicentra cucullaria* (L.),

Dutchman's Breeches; b, *Corydalis aurea*, Willd. Golden Corydalis.

This is only a small family. The old-fashioned favorite "Bleeding Heart" belongs here.

51b Stamens not united; flowers regular; sepals 4. Figs. 359 and 360. 52

52a Stamens alike; leaves usually palmately compound; capsule 1-celled.　　　　　Caper Family, CAPPARIDACEAE

Figure 359

Fig. 359. *Cleome spinosa* L., Giant Spider Plant.

Though this is only a weed introduced from tropical America, it is often planted as an ornamental garden plant. A few trees belong to this family.

52b Stamens in two whorls, 4 long and 2 short (rarely only 2 or 4); capsule 2-celled. Mustard Family, CRUCIFERAE

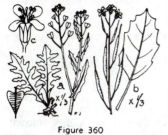

Fig. 360. a, *Capsella bursa-pastoris* (L.), Shepherd's Purse; b, *Brassica juncea* (L.), Indian Mustard; c, typical mustard flower.

Common examples: Turnip, Rape, Kohlrabi, Cauliflower, Cabbage, Horse Radish, Honesty, and Radish.

Figure 360

53a Stamens as many as the petals with a stamen in front of each petal.

Fig. 361a. 54

53b Stamens between the petals (alternating) or more numerous. Figs. 361b and c. 57

Figure 361

54a Calyx of 2 sepals; flowers perfect.
Purslane Family, PORTULACACEAE

Fig. 362. *Claytonia virginica* L., Spring Beauty.

This plain-looking little plant literally carpets the woods throughout much of the East and South in early spring. The flowers are white with a faint pinkish tint.

Figure 362

54b Calyx with more than 2 sepals; fruit a drupe, berry or capsule. Figs. 363 to 365. 55

55a Petals 6 or more; petals and sepals both imbricated in the bud. Ovary 1-celled.
Barberry Family, BERBERIDACEAE

Fig. 363. *Jeffersonia diphyllum* (L.), Twin Leaf.
The genus name was given in honor of Thomas Jefferson; the species name refers to the divided leaves. The flowers are white.
The family is named from the barberry shrub, one species of which is the alternating host of wheat rust.

Figure 363

**55b Petals and stamens only 4 or 5; ovary 2-4 celled.
Figs. 364 and 365.** **56**

**56a Tendril-climbing woody vines, rarely shrubs; petals falling very
early, calyx minute; fruit a berry.**

Grape Family, VITACEAE

Fig. 364. a, *Psedera quinquefolia* Gr., Virginia Creeper, a highly attractive *non-poisonous "Ivy"*; b, *Vitis vulpina* L., River Grape.

Wild grapes are widely distributed. From them many of our best cultivated varieties have been developed.

Figure 364

**56b Shrubs or small trees, rarely vines; fruit a drupe or capsule; calyx
plainly 4 or 5-parted; petals sometimes wanting.**

Buckthorn Family, RHAMNACEAE

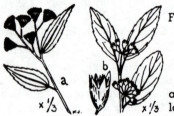

Fig. 365. a, *Ceanothus americanus* L., New Jersey Tea. The small white flowers are attractive. b, *Rhamnus caroliniana* Walt., Carolina Buckthorn.

The flowers of the New Jersey Tea are white and very soft. They attract large numbers of insects so that the insect collector does well to know this plant.

Figure 365

57a Ovary with only 1 cell. Figs. 366 to 374. . . . . **58**

57b Ovary with 2 or more cells. Figs. 375 to 402. . . . . **65**

**58a (a, b, c) Ovules or seeds attached to walls of
ovary (parietal placenta). Fig. 366a.** **59**

**58b Ovules or seeds attached at base or center
of ovary (central placenta). Fig. 366b.** **64**

Figure 366

58c Usually but one ovule in a one-celled ovary (rarely more than 1-cell but then with one ovule to each cell). Trees or shrubs with resinous bark or milky sap.

Cashew Family, **ANACARDIACEAE**

Figure 367

Fig. 367. a, *Rhus toxicodendron* L., Poison Ivy; b, *Rhus glabra* L., Smooth Sumac.

It is of vital importance for every nature lover to recognize Poison Ivy.

Several of the Sumacs are used for landscaping.

59a Stamens with filaments united. See Fig. 352.

Mallow Family, **MALVACEAE**

59b Stamens on separate filaments. Figs. 368 to 372. . . . .**60**

60a Fertile stamens 5; with numerous sterile stamens (staminodia) at base of each petal. Grass-of-Parnassus Family, **PARNASSIACEAE**

× ¼

Figure 368

Fig. 368. *Parnassia caroliniana* Michx., Grass-of-Parnassus.

In bogs and moist soil. Flowers greenish white. This is a comparatively rare and little known plant.

60b All stamens fertile. Figs. 369 to 372. . . . .**61**

61a Leaves with gland-tipped sticky hairs for catching insects.

Sundew Family, **DROSERACEAE**

× ⅓

Figure 369

Fig. 369. *Drosera rotundifolia* L., Round-leaved Sundew.

In bogs, widely distributed. These small fragile plants are likely to prove disappointing to one who has studied about them, when they are seen, for the first time. The insects they catch are usually tiny ones. There are several species.

61b Leaves not for catching insects. Figs. 370 to 372. 62

62a Leaves with black or transparent dots; entire or scale-like.
St. John's-wort Family, **HYPERICACEAE**

Fig. 370. a, *Triadenum virginicum* L., Marsh St. John's-wort. Flowers reddish-purple; stamens in 3 sets of 3 each; b, *Sarothra gentianoides* L., Orange Grass. Grows in sandy soil.

Figure 370

62b Leaves not dotted. Figs. 371 and 372. 63

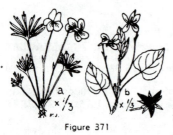

63a Corolla irregular; flowers pansy-shaped. Violet Family, VIOLACEAE

Fig. 37. a, *Viola pedata* L., Bird'sfoot Violet; b, *Viola striata* Ait., Pale Violet. Violets are favorites everywhere. The pansy belongs to this same family. There are many species and they are widely distributed.

Figure 371

63b Corolla regular; sepals and petals 3 or 5.
Rock-rose Family, **CISTACEAE**

Fig. 372. *Lecha tenuifolia* Michx., Narrow-leaved Pin-weed.

The flowers are a purplish-red. The species is widely scattered east of the Missouri River.

Figure 372

64a Herbs with opposite or whorled, usually entire leaves; stems usually swollen at the joints. Pink Family, **CARYOPHYLLACEAE**

Fig. 373. a, *Cerastium nutans* Raf., Nodding Chickweed; b, *Saponaria officinalis* L., Bouncing Bet.

This rather large family includes the Carnation, Sweet William, Baby's Breath, and Garden Pinks.

Figure 373

64b Shrubs with alternate leaves usually scale-like.
Tamarisk Family, TAMARICACEAE

Fig. 374. *Tamarix gallica* L.

Though frequently planted, these scraggly, open shrubs are curious rather than attractive. The flowers are pink. It becomes a sizable tree in the South-west.

Four genera and almost a hundred species make up the family. They belong to the Eastern Hemisphere.

Figure 374

65a Stamens united with each other and with the thickened stigma arising from the two ovaries. Milkweed Family, ASCLEPIADACEAE

Fig. 375. a, *Asclepias incarnata* L., Swamp Milkweed; b, *Acerates viridiflora* (Raf.), Green Milkweed.

There are many very attractively marked and colored milkweeds. They are widely distributed and vary much in size.

Figure 375

65b Stamens not as in 65a. Figs. 376 to 402. ....**66**

66a Flowers irregular (bilaterally symmetrical).
Figs. 376 to 379. ....**67**

66b Flowers regular (radially symmetrical).
Figs. 380 to 402. ....**69**

67a Trees or shrubs; leaves palmately compound, opposite.
Soapberry Family, SAPINDACEAE

Fig. 376. *Aesculus glabra* Willd., Ohio Buckeye.

This family includes a number of foreign plants cultivated as ornamentals. Buckeyes and Horsechestnuts are widely distributed. There are several species, some with red flowers.

Figure 376

67b Herbs; leaves simple. Figs. 377 to 379. 68

68a (a, b, c) Ovary 5-celled; stems soft, succulent, stamens 5-10; one sepal prolonged back into nectar sac or spur.

Balsam Family, BALSAMINACEAE

Fig. 377. *Impatiens biflora* Walt., Wild Touch-me-not. Flowers vermilion. The Pale Touch-me-not is a larger plant with pale yellow flowers. It is the seed pods (b) that may "not-be-touched" for they "explode" with the slightest irritation.

Figure 377

68b Ovary 3-celled; stamens 8, leaves peltate or palmately divided.
Nasturtium Family, TROPAEOLACEAE

Fig. 378. *Tropaeolum majus* L., Garden Nasturtium.

This favorite flower-garden ornamental is a native of South America. The stems, seeds and buds are sometimes used in pickles for their flavor.

Figure 378

68c Ovary 2-celled; stamens 6-8; leaves simple, entire, with stipules.
Milkwort Family, POLYGALACEAE

Fig. 379. a, *Polygala viridescens* L., Purple Milkwort; b, *Polygala polygama* Walt. Racemed Milkwort.

The family has about 1000 species, many of which are confined to the tropics. This genus alone has over 400 species.

69a Stamens same number as the petals or twice as many.
Figs. 380 to 398. 70

69b Stamens not as in 69a. Figs. 399 to 402. 89

73a Leaves opposite; lobed and palmately-veined or pinnately-compound; no stipules. Fruit splitting into two-winged samaras.

Maple Family, ACERACEAE

Fig. 380. *Acer platanoides* L., Norway Maple.

Sugar-making from maple sap was once a highly important early spring industry but is losing its importance. Maple trees furnish some fine lumber.

Figure 380

74a Leaves opposite; woody veins, shrubs or small trees.

Staff-tree Family, CELASTRACEAE

Fig. 381. *Euonymus atropurpureus* Jacq., Eastern Wahoo.

The salmon-pink fruit which upon opening displays its vermilion seeds never fails to attract attention. There are in all some 65 species, some of which are evergreens.

Figure 381

75a Climbing woody vines.   Staff-tree Family, CELASTRACEAE

Fig. 382. *Celastrus scandens* L., Bittersweet.

Its fruit is much prized for winter decoration. In color it is much like the Wahoo. This genus has some 30 species.

Figure 382

75b Woody plants, not climbing. Figs. 383 and 384. 76

76a Flowers small, in racemes. Leaves thick, entire. Fruit dry.

Cyrilla Family, CYRILLACEAE

Figure 383

Fig. 383. *Cyrilla racemiflora* L., Leatherwood.

This small family of trees and shrubs belongs in the South. The species here used as an example grows in wet places.

76b Flowers solitary or clustered in the axils. Leaves often leathery, simple; fruit a berry-like drupe with several hard seeds.

Holly Family, AQUIFOLIACEAE

Figure 384

Fig. 384. *Nemopanthus mucronata* (L.), Mountain Holly.

It is found in swamps. It is not evergreen as most of the other members of the family.

77a Leaves with glandular punctations (translucent dots). Pistils sometimes 1-5, distinct.

Rue Family, RUTACEAE

Figure 385

Fig. 385. a, *Ptelea trifoliata* L., Common Hoptree; b, *Zanthoxylum americanum* Mill., Common Prickly-ash.

The Hoptree, a shrub or small tree, has very characteristic fruit which has won for it the name "Wafer-Ash". Its range is throughout much of the eastern half of the U. S.

77b Leaves not punctate. Figs. 386 and 387. 78

78a Fruit (in our one species) a winged samara. An open-growing tree with thick branches and long pinnately-compound leaves.

Ailanthus Family, SIMARUBACEAE

Figure 386

Fig. 386. *Ailanthus altissima* Swingle, Tree-ofheaven.

This tree is a native of China. It grows rapidly and its leaves have an unpleasant odor. It sprouts vigorously and at considerable distances from the parent tree.

78b Fruit of many forms but not a samara.

Soapberry Family, SAPINDACEAE

Figure 387

Fig. 387. *Sapindus drummondii* H. & A.; Western Soapberry.

This tree may attain a height of 50 feet. It belongs to the South and West.

79a Flowers imperfect; monoecious or dioecious; sap usually acrid or milky.

Spurge Family, EUPHORBIACEAE

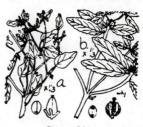

Figure 388

Fig. 388. a, *Croton monanthogynus* Michx., Single Fruited Croton; b, *Croton glandulosa* L. Glandular Croton.

This is a large important family. With many species the maturing ovary is so large and so suspended on a slender stem that it hangs out of the flower in a very characteristic way.

Common examples: Poinsettia, Castor-oil Plant, Para Rubber Tree.

80b Twice as many carpels as sepals. Figs. 391 and 392. 82

81a One ovule to each carpel; ovary 2-3 celled.

False Mermaid Family, LIMNANTHACEAE

Fig. 389. *Floerkea proserpinacoides* Willd., False Mermaid.

Flowers red, pink or white. The species used here grows in marshes and other wet places.

Figure 389

81b One or two ovules to each carpel; ovary 5-celled, splitting at maturity; stamens with versatile anthers.

Geranium Family, GERANIACEAE

Fig. 390. a, *Geranium maculatum* L., Wild Crane's Bill; b, *Erodium circutarium* (L.), Red-stem Filaree.

The showy cultivated "geraniums" belong to this family, but to the genus *Pelargonium* rather than *Geranium*.

Figure 390

82a Leaves pinnately compound. Caltrop Family, ZYGOPHYLLACEAE

Fig. 391. *Kallstroemia intermedia* Rydb.; Greater Caltrop.

The flowers are yellow and the plants hairy.
Lignum-vitae, a tropical tree with unusually hard heavy wood belongs here.

Figure 391

82b Leaves simple, stipules small or none, syles 2-5.

Flax Family, LINACEAE

Fig. 392. a, *Linum usitatissimum* L., Flax.

This important plant is widely cultivated. Its fibers produce linen and from its seed comes linseed oil. The flowers are sky-blue. A large field, when in bloom, makes a sight long to remember.

Figure 392

83a Leaves simple. Figs. 393 to 396. 84
83b Leaves compound. Figs. 397 and 398. 88
84a Leaves simple, opposite, with stipules between them; small plants
    growing in marshes.   Waterwort Family, ELATINACEAE

Fig. 393. *Elatine triandra* Schk., Waterwort.

    This is a mid-west and western plant that lives, as its name indicates, in lakes and such. The family has about 30 widely distributed species.

Figure 393

84b Leaves alternate, or if opposite then without stipules.
    Figs. 394 to 398. 85
85a Pistil with 2 to 5 styles. 87
85b Pistil with but one style. Figs. 394 and 395. 86
86a Stamens arising from the calyx; leaves usually opposite.
                   Loosestrife Family, LYTHRACEAE

Fig. 394. *Lythrum alatum* Pursh., Wing-angled Loosestrife.

    It grows in low damp ground; its flowers are purple. The very showy shrub, Crape-Myrtle, of our southern states, belongs in this family.

Figure 394

86b Stamens not attached to the calyx.   Heath Family, ERICACEAE

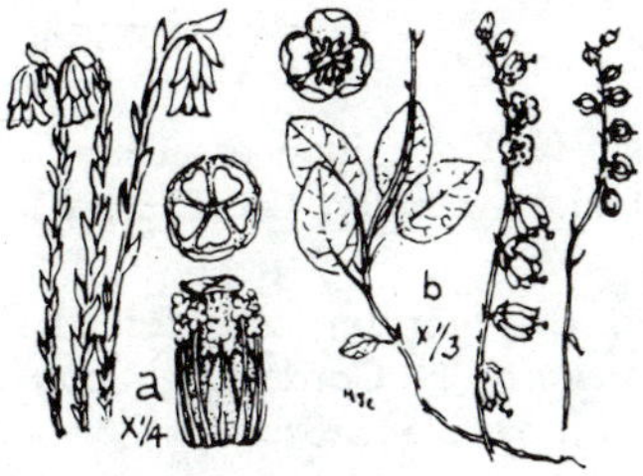

Fig. 395. a, *Monotropa uniflora* L., Indian Pipe; b, *Pyrola secunda* L., One-sided Wintergreen.

    The family is very important in the decorative plants both native and cultivated, that belong to it, such as the Heaths, Heather, Laurels, Rhododendrons, and Azaleas.

Figure 395

87a Stamens arising from the calyx. Fig. 395.
                   Heath Family, ERICACEAE

87b Stamens not attached to calyx; nodes of stems usually swollen.
Pink Family, CARYOPHYLLACEAE

Figure 396

Fig. 396. *Silene noctiflora* L., Night-flowering Catchfly.

This family while producing some valued ornamentals, also has many serious weeds among its members.

Figure 396

88a Herbs with sour sap; ovary 5-celled; stamens 10-15; leaves usually with 3 obcordate leaflets.
Wood Sorrel Family, OXALIDACEAE

Fig. 397. *Oxalis corniculata* (L.), Yellow Woodsorrel.

These plants are often called "Sheep-sorrel" but that name should be reserved for *Rumex acetosella*.

Figure 397

88b Shrubs or trees; carpels usually 3; fruit an inflated capsule.
Bladdernut Family, STAPHYLEACEAE

Fig. 398. *Staphylea trifolia* L., American Bladdernut.

This is a small family with several species being raised for ornament. The inflated fruit never fails to attract attention.

Figure 398

90a Sepals and petals 4; stamens 6 or less; fruit a silique or silicle.
Mustard Family, CRUCIFERAE

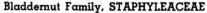

Fig. 399. a, *Sisymbrium officinale* L., Hedge Mustard; b, *Cardamine douglassii* (Torr.), Purple Cress.

The cross-shaped, 4-petaled flowers are characteristic of this important family as are also the seed pods.

Figure 399

90b Sepals and petals 5.   St. John's-wort Family, HYPERICACEAE

Figure 400

Fig. 400. *Hypericum mutilum* L., Dwarf St. John's-wort.

Most of the members of the family are characterized by their many stamens; this species is one of the exceptions.

91a Petals 4, stamens fewer.   Olive Family, OLEACEAE

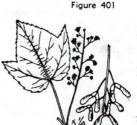

Figure 401

Fig. 401. a, *Ligustrum vulgare* L., Privet; b, *Chionanthus virginica* L., Fringetree.

As the family name indicates, the Olive belongs here. Other important members of the family are Ash, furnishing excellent wood for tool-making, and Lilac and Jasmine.

91b With more stamens than petals.
Maple Family, ACERACEAE

Figure 402

Fig. 402. *Acer spicatum* Lam., Mountain Maple.

The family boasts but two genera but many species of widely distributed trees and shrubs. Maples are important for ornament, their wood and their sugar.

92a (a, b, c) Herbaceous vines, bearing tendrils.
Gourd Family, CUCURBITACEAE

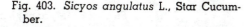

Fig. 403. *Sicyos angulatus* L., Star Cucumber.

This important family includes many food plants; Pumpkins, Squashes, Gourds, Watermelons, Muskmelon, Cucumbers, and some ornamental vines.

Figure 403

92b Exotic, usually evergreen aromatic shrubs or trees; sepals and petals 4-5, stamens many, ovary 1 to many-celled.

Myrtle Family, **MYRTACEAE**

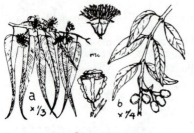

Fig. 404. a, *Eucalyptus* sp.; b, *Eugenia myrtifolia* Sims., Australian Bush-Cherry.

This important family belongs to Australia and the tropics. It yields valuable lumber, oils, gums, cloves, allspice and many other commercial products.

Figure 404

92c Not as in 92a or 92b. Figs. 405 to 417. 93

93a But one seed or ovule in each cell of ovary.
Figs. 405 to 410. 94

93b More than one seed or ovule in each cell of the ovary.
Figs. 411 to 417. 99

94a With 2, 4, or 8 stamens. Figs. 405 to 407. 95

94b With 5 or 10 stamens. Figs. 408 to 410. 97

95a Shrubs and trees with drupe-like fruit and single style and stigma.

Dogwood Family, **CORNACEAE**

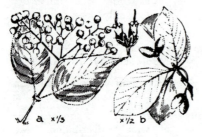

Fig. 405. a, *Cornus drummondi* Meyer; Roughleaf Cornel; b, *Nyssa sylvatica biflora* Sarg. Swamp Tupelo.

The family consists mostly of shrubs and trees though an occasional herb figures in it. There are some 10 genera in all but *Cornus* is the most important.

Figure 405

95b Herbs. Figs. 406 and 407. 96

96a Styles 4 (sometimes with 4 sessile stigmas).
Water-milfoil Family, **HALORAGIDACEAE**

Fig. 406. *Myriophyllum spicatum* L., Spiked Water-milfoil.

These watter plants play an important part in our watercourses in fish and game culture. Some species are seen in indoor aquaria.

Figure 406

96b With but one style; stigma 2 to 4-branched.
Evening-Primrose Family, ONAGRACEAE

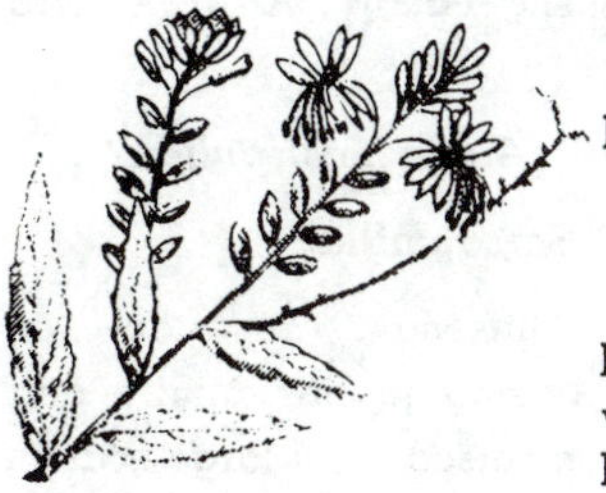

Fig. 407. *Gaura biennis* L., Biennial Gaura.

The ovary develops into a ribbed achene.

Tropical America and New Zealand brings us the Fuchsia, several species of which are common ornamentals. They belong to this family.

Figure 407

97a Fruit fleshy. Figs. 409 and 410. 98

97b Fruit dry when ripe; herbs; flowers small, generally in simple or compound umbels; petals 5; stamens 5; leaves alternate and usually compound.
Carrot Family, UMBELLIFERAE

Fig. 408. a, *Pastinaca sativa* L., Parsnip; b, *Eryngium yuccaefolium* Mich., Rattlesnakemaster.

This large and distinctly marked family includes many favorite ornamentals as well as some well-known food plants, such as Carrot, Parsley, and Celery. Herbs for seasoning furnished by the family include Dill, Cumin, Coriander, Lovage, Anise, Myrrh, and Caraway. Several highly poisonous plants also belong here, the Poison-Hemlock being an outstanding example.

Figure 408

98a Leaves simple; with no prickles. Trees or shrubs. Fruit a pome.
Rose Family (in part), ROSACEAE

Fig. 409. *Crataegus punctata* Jacq., Dotted Thorn.

This is an exceptionally large family. The genus *Cratatgus* in itself has more than a thousand species. These Hawthorns or Red Haws are widely distributed trees and shrubs. They are beautiful both in flower and in fruit.

Figure 409

133

98b Leaves compound; or if simple, with prickles. Herbs, shrubs, trees. Fruit a drupe or berry.　　　　　　Ginseng Family, ARALIACEAE

Figure 410

Fig. 410. *Aralia nudicaulis* L.,

Wild Sarsaparilla.

The Ginsengs, important as drugs, belong to this family and are often raised on a large scale.

99a Spiny, fleshy plants; stems often jointed; leaves absent or small; numerous petals and sepals; stamens on a hypanthium.
　　　　　　　　　　　　　　Cactus Family, CACTACEAE

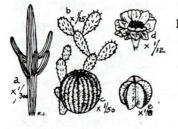

Figure 411

Fig. 411. a, *Carnegiea gingantea* Suwarro, Giant Cactus, the largest known cactus; may attain a height of 40 feet; b, *Opuntia engelmannii* S-D.; c, *Echinocactus grusonii* Hild, Golden Cactus; d, typical cactus flower; e, *Astrophytum myriostigma*, Bishop's Cap.

99b Neither spiny nor fleshy; often woody plants; petals and sepals normally 5; stamens numerous on a hypanthium.
　　　　　　　　　　　　　　Rose Family, ROSACEAE

Fig. 411½. *Malus pumila* Mill. Apple. This is one of the largest and most important families of plants. Many of our most prized fruits belong to it.

Common examples of the family: Pears, Quinces, Crab Apples, Raspberries, Dewberries, Loganberries, Strawberries, Peaches, Apricots, Plums, Cherries, etc.

Figure 411½

101a Sepals or calyx lobes 2; smooth herbs with fleshy entire leaves.

Purslane Family, PORTULACACEAE

Fig. 412. *Portulaca oleraceae* L., Purslane.

Very common in gardens. Being a xerophyte, it can stand much more abuse than most of its competitors.

Figure 412

101b Sepals or calyx lobes more than 2. Figs. 413 and 414. 102
102a Erect or climbing herbs, with stinging or glandular hairs; stamens many; petals 5 or sometimes appearing as 10.

Loasa Family, LOASACEAE

Fig. 413. a, *Mentzelia albicaulis* Dougl., White-stemmed Mentzelia; b, *Nuttallia decapetala* (Pursh), Prairie Lily.
This family belongs principally to the western part of our country. The greater number of species belong in South America.

Figure 413

102b Shrubs with small solitary or racemed flowers. Petals or stamens 4 or 5.

Saxifrage Family, SAXIFRAGACEAE

Fig. 414. *Ribes gracile* Pursh., Missouri Gooseberry.

Gooseberries and Currants are western hemisphere plants. Our cultivated forms have been bred up from the native stock. The Golden Currant is an ornamental shrub.

Figure 414

103a Anthers opening by pores at their tip. Leaves opposite with 3 to 9 nerves.
Meadow-beauty Family, MELASTOMACEAE

Fig. 415. *Rhexia virginica* L., Meadow-beauty.

This large family is pretty much confined to the tropics. Some of these tropical species may be found in hot-houses.

Figure 415

103b Anthers opening along their sides; stamens arising from the calyx. Figs. 416 and 417. 104

104a Stamens 4 or 8; styles 1; base of calyx usually forming an elongated tube.

Evening-Primrose Family, ONAGRACEAE

Fig. 416. Oenothera biennis L., Common Evening-Primrose.

If one wishes to see flowers open naturally, this family and the above species makes an excellent example. The whole process, once it has started, takes but a few minutes.

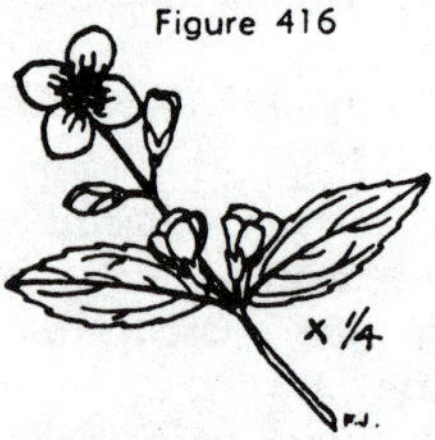

Figure 416

104b Styles 2 to 5; stamens 8 to many.

Saxifrage Family, SAXIFRAGACEAE

Fig. 417. Philadelphus cornarius, Mock Orange. This species can be told from the other Syringas by the fragrant odor of its flowers.

Figure 417

105a Calyx arising below the ovary (or surrounding its base). (Ovary superior). Fig. 418a. 113

105b Calyx arising above the ovary. (Ovary inferior). Fig. 418b. 106

Figure 418

106a Flowers of one or more kinds crowded in heads with one or more rings of bracts beneath; the 5 anthers united at their sides to form a tube surrounding the style (when present).

Composite Family, COMPOSITAE

Fig. 419. a, Taraxacum officinale Weber, Common Dandelion; b, Ambrosia artemisiaefolia L., Common Ragweed; c, Erigeron caespitosus Nutt., Tufted Erigeron; d, Liatris cylindracea (Michx.), Cylindric Blazing Star.

This great family is readily distinguished by the characters given in the key. Many favorite ornamentals belong to it

Figure 419

such as Sunflowers, Asters, Daisies, Marigolds, Strawflowers, Zinnias, Dahlias and Chrysanthemums. The family also furnishes a few food plants, Artichoke, Vegetable Oyster and Chicory.

106b Flowers not in dense-bracted (involucrate) heads. Figs. 420 to 426. 107

107a Herbaceous tendril-bearing vines. Fruit a pepo (pumpkin-like). Leaves palmately compound, lobed or veined.

Gourd Family, CUCURBITACEAE

Fig. 420. *Cucumis sativis* L., Cucumber.

The flowers of many of the species are very conspicuous yet they are little used for ornament. Gourds are prized for their decorative fruit.

Figure 420

107b Without tendrils. Figs. 421 to 426. 108

108a Stamens united by their anthers; flowers irregular usually in racemes, never in an involucrate head.

Lobelia Family, LOBELIACEAE

Fig. 421. *Lobelia cardinalis* L., Cardinalflower.

This is likely the most decorative species of the family. It grows in damp places and is very showy with its brilliant red flowers.

The members of this family are easily recognized. There are some 600 species in all.

Figure 421

108b Stamens not united. Figs. 422 to 426. 109

109a Stamens arising from the walls of the corolla. Figs. 423 to 426. 110

109b Stamens not as in 109a; juice milky.

Blue-bell Family, CAMPANULACEAE

Fig. 422. a, *Campanula rotundifolia* L., Blue Bells; b, *Specularia perfoliata* L., Venus' Looking-glass. Harebells, Canterbury Bells, and

Balloon-Flowers, among other favorite cultigens belong here. Blue is the prevailing color of the flowers.

Figure 422

110a Stamens 1 to 3 (rarely 4), fewer than the lobes of the corolla.
Valerian Family, VALERIANACEAE

Figure 423

Fig. 423. *Valeriana officinalis* L., Garden Heliotrope.

This European plant is sometimes found growing wild. It is often seen in gardens. There are 300 widely scattered species.

110b Stamens 4-5; leaves opposite or whorled. Figs. 424 to 426. ...111

111a Ovary with but one cell. Flowers in dense involucrate heads.
Teasel Family, DIPSACACEAE

Figure 424

Fig. 424. *Dipsacus sylvestris* Huds., Common Teasel.

Another European plant introduced here as an ornamental and for use in weaving wool. The Pincushion Flower, a common garden favorite also belongs here.

111b Ovary with more than 1 cell. Figs. 425 and 426. 112

112a Leaves whorled or opposite and with stipules; petals usually 4.
Madder Family, RUBIACEAE

Figure 425

Fig. 425. a, *Houstonia coerulea* L., Bluets; b, *Galium aparine* L., Bedstraw.

From the lowly and despised Bedstraw to the prized Gardenia or the commercially important Coffee seems a far cry but all belong to this rather large family. The family also yields medical herbs and dye stuffs.

138

112b Leaves opposite, without stipules; often perfoliate; flowers frequently irregular. Honeysuckle Family, CAPRIFOLIACEAE

Fig. 426. a, *Lonicera japonica* Thunb., Chinese Honeysuckle; b, *Sambucus canadensis* L., American Elder.

Honeysuckles may usually be told by the flowers and fruit growing in pairs. The leaves are often perfoliate. Buckbrush or Coral-berry prized by landscapists as a hedge shrub but a serious annoyance in farm pastures is a member of the family.

Figure 426

113a Stamens more in number than the lobes of the corolla. Figs. 427 to 436. 114
113b Stamens not exceeding the number of corolla lobes. Figs. 437 to 472. 122
114a Ovary with but one cell. Figs. 427 to 430. 115
114b Ovary with more than one cell. Figs. 431 to 436. 116
115a (a, b, c, d) Ovules (seeds) all attached to one side of ovary (pod). Flowers often pea-shaped although sometimes nearly regular. Leaves usually compound with stipules.

Pea Family, LEGUMINOSAE

Fig. 427. a, *Prosopia glandulosa* Torr., Prairie Mesquite; b, *Gymnocladus dioica* (L.), Kentucky Coffeetree; c, *Robinia hispida* L., Rose Acacia.

The outstanding valuable feature about the family is its use as a soil builder. Nitrogen-fixing bacteria which live in nodules on the roots of the legumes have the ability to take nitrogen from the air and to make nitrogenous compounds for these bacteria and their host.

Figure 427

115b Ovules attached in two rows on opposite sides of ovary or pod. Flowers irregular. Sepals 2. Petals 4.

Fumitory Family, FUMARIACEAE

Fig. 428. *Adlumia fungosa* (Ait.), Climbing Fumitory.

The flowers of this species are greenish purple. The sepals of all members of this family are small and fall early. There are two pairs of petals which differ markedly from each other.

Figure 428

115c Ovules (seeds) attached at center or base of ovary. Flowers regular. Woody plants. Storax Family, STYRACACEAE

Fig. 429. *Styrax americana* Lam., Smooth Storax.

Most of the members of this small family of trees and shrubs are tropical.

Opossum-wood, a beautiful shrub or small tree belongs to this family.

Figure 429

115d Ovules many, attached to five ridges (placentae) on the walls of the fleshy, melon-like fruit. Petals on the carpelate flowers almost distinct. Small somewhat palm-like trees; tropical.
Pawpaw Family, CARICACEAE

Fig. 430. *Carica papaya* L., Papaya.

This plant must be raised where there is no danger of frost. It is highly regarded for its fruit. It is the source of the drug papain.

Figure 430

116a Ovary with but 2 seeds and 2 cells. Flowers irregular. Stamens usually 8.   **Milkwort Family, POLYGALACEAE**

Fig. 431. *Polygala alba* Nutt., White Milkwort.

This species is a prairie plant of our western states. The flowers are white.

This large family contains some 1000 species.

Figure 431

116b Ovary with 3 or more cells. Figs. 432 to 436. 117

117a Plants without green coloring matter; living as saprophytes on humus or decaying roots, etc.   Heath Family, ERICACEAE

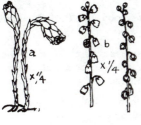

Figure 432

Fig. 432. a, *Monotropa uniflora* L., Indian Pipe; b, *Pterospora andromedea* Nutt., Pine Drops.

Someone told us recently of their unsuccessful attempts to transplant Indian Pipe. It could doubtless be done but the decaying organic matter on which it was feeding would need to be brought along also. To raise these plants from seed would likely be easier but even then proper food would be necessary.

117b Plants with green leaves. Figs. 352 and 433 to 436. 118

118a Herbs; stamens united by their filaments into a tube. See Fig. 352.   Mallow Family, MALVACEAE

118b Woody plants. Figs. 433 to 436. 119

119a Stamens not united by their filaments. Figs. 433 and 434. 120

119b Stamens united by their filaments into one or more groups. Figs. 435 and 436. 121

Figure 433

120a Styles 4. Calyx 3 to 7 lobed. Filaments short.   Ebony Family, EBENACEAE

Fig. 433. *Diospyros virginiana* L., Common Persimmon.

The fruit of the persimmon is prized for food when thoroughly ripe. It is so bitter as to be long remembered if eaten when green. The wood is unusually hard. The best ebony comes from the heartwood of *Diospyros ebenum*, a native of India and Ceylon.

120b Ovary with but one style. Calyx 4 or 5 lobed.
**   Heath Family, ERICACEAE**

Figure 434

Fig. 434. a, *Epigaea repens* L., Trailing Arbutus; b, *Vaccinium corymbosum* L., Tall Blueberry.

Surely a lot of sentiment centers around the Trailing Arbutus and who does not enjoy a piece of blueberry or huckleberry pie. "Pretty slow work picking them".

141

121a Ovary wholly superior. Stamens numerous.

Tea Family, THEACEAE

Figure 435

Fig. 435. *Stewartia pentagyna* L'Her., Mountain Stewartia.

Tea is made from the leaves of *Thea sinensis* L. of this family. It is a shrub or small tree and is native of China and India.

Its cultivation in these countries has long been an important industry.

121b Ovary at most but partly superior.

Storax Family, STYRACACEAE

Figure 436

Fig. 436. *Halesia carolina* Ellis, Carolina Silverbell

A small tree with white flowers. Grows in woods throughout much of our southeast. The "bells" are considerably larger on the plants growing in the Blue Ridge mountains. It is named *H. monticola* Sarg.

122a Stamens as many as the corolla lobes and directly in front (opposite) of them; corolla lobes all alike. Figs. 437 to 439. 123

122b Stamens between the corolla lobes (alternating with them) or fewer than the lobes. Figs. 440 to 472. 125

123a Styles 5, separate or united; fruit dry.

Plumbago Family, PLUMBAGINACEAE

Figure 437

Fig. 437. a, *Statica armeria* L., Sea Pink; b, *Plumbago capensis* Thunb., Leadwort.

The Leadwort is a delicate but highly decorative plant used much in the South. Its flowers are pale blue or sometimes white. It is a native of Africa. A red-flowered species comes from Asia.

142

123b Style 1. Figs. 438 and 439. 124

124a Trees or shrubs, often with milky sap.
Sapodilla Family, SAPOTACEAE

Fig. 438. *Bumelia lycoides* (L.), Buckthorn Bumelia.

The family is a small one and largely tropical. A number of these exotic species are raised for ornament.

The Sapodilla a tropical tree raised in our southern areas bears delicious fruit.

Figure 438

124b Herbs; fruit a one-celled capsule with few to many seeds.
Primrose Family, PRIMULACEAE

Fig. 439. a, *Dodocatheon meadia* L., Shooting Star; b, *Steironema ciliatum* (L.), Fringed Loosestrife.

This family produces several important house plants. It should not be confused with the Evening Primrose family for they are quite different plants.

Figure 439

125a Corolla regular (the lobes all alike). Figs. 440 to 462. 126

125b Corolla irregular (the lobes not all alike). Figs. 463 to 472. 145

126a Stamens the same number as the lobes of the corolla.
Figs. 440 to 458. 127

126b Stamens fewer than the lobes of the corolla. Figs. 459 to 462...142

127a (a, b, c) Ovary 1, deeply four lobed. Fig. 440. ..129

Figure 440

127b Ovary 1, not deeply lobed. Figs. 445 to 458. 130

127c Ovaries 2 or sometimes one with two horns; sap usually milky.
Figs. 441 and 442. 128

128a Stamens united with each other and with the stigma; styles distinct.   Milkweed Family, ASCLEPIADACEAE

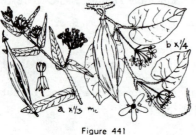

Figure 441

Fig. 441. a, *Asclepias tuberosa* L., Butterfly-weed; b, *Vinvetoxicum gonocarpos* Walt., Large-leaved Angle-pod.

It is the structure of the flower and fruit and not the milky sap that makes a plant a Milkweed. The some 2000 species are scattered pretty much the world over.

128b Stamens not united; no stipules.

Dogbane Family, APOCYNACEAE

Figure 442

Fig. 442. a, *Apocynum androsaemif o l i u m* L., Spreading Dogbane; b, *Vinca minor* L., Periwinkle.

The Oleander, a very showy shrub or tree, native of Asia, falls here. They (there are 3 species) are raised in the greenhouse in the north but grow out of doors farther south.

129a Leaves alternate; flowers often blue though sometimes yellow or other colors.   Borage Family, BORAGINACEAE

Figure 443

Fig. 443. a, *Mertensia virginica* (L.), Virginia Cowslip; b, *Lithospermum canescens* (Michx.), Hoary Puccoon.

The family is remembered for its "Forget-me-nots" with their pale blue flowers with white eyes. The plants of this family are usually hairy and the flowers often brilliantly colored.

129b Leaves opposite; stems usually square in cross section.

Mint Family, LABIATAE

Figure 444

Fig. 444. *Isanthus brachiatus* (L.), False Pennyroyal.

Many ornamentals, and some herbs used for seasoning, fall here. Of the latter, Garden Sage, Thyme, Peppermint, Spearmint, Pennyroyal (some rival gum maker should try it; it's a good one), Hoarhound, and Summer Savory are examples.

Figure 445

131a Leaves entire and opposite. (Ovary sometimes partly 2-celled).

Gentian Family, GENTIANACEAE

Fig. 445. a, *Gentiana andrewsii* Griseb., Closed Gentian; b, *Sabbatia angularis* (L.), Rose Pink.

The Fringed Gentian is perhaps the best known from books. All of these plants are comparatively rare.

The family includes more than 700 species.

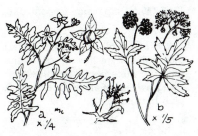

Figure 446

131b Leaves lobed or toothed or compound.

Water-leaf Family, HYDROPHYLLACEAE

Fig. 446. a, *Ellisia nyctelea* L., Nyctelea; b, *Hydrophyllum virginicum* L., Virginia Water-leaf.

This family belongs quite largely to Western North America. *Ellisia* is a very common weed in gardens and neglected spots.

Figure 447

132a A yellow leafless thread-like plant twining around other plants on which it lives parasitically.

Dodder Family, CUSCUTACEAE

Fig. 447. *Cuscuta gronovii* Willd., Lovevine.

These plants have no chlorophyll and after they contact their host plant abandon their roots. About 100 species are known. Parasites among the flowering plants are not common.

132b Not parasitic. Figs. 448 to 458. 133

133a Leaves opposite and stipulate or their bases connected by stipulate lines.

Logania Family, LOGANIACEAE

Figure 448

Fig. 448. *Gelsemium sempervirens* (L.), Yellow Jessamine.

An Indian tree of this family produces strychnine. The seeds are used for this purpose. The family is mostly tropical, and numbers some 400 species.

133b Leaves alternate or if opposite without stipules or stipular lines. Figs. 449 to 458. 134

134a (a, b, c) Stamens not attached to corolla or scarcely so. Figs. 450 and 451. 135

134b Stamens in the notches of the corolla. Low shrubs or herbs.

Diapensia Family, DIAPENSIACEAE

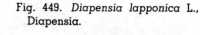

Fig. 449. *Diapensia lapponica* L., Diapensia.

It grows well up on our eastern mountains. The flowers are white. Other members of this little family grow in open woods and in pine barrens.

Figure 449

134c Stamens attached to the walls of the corolla. Figs. 452 to 458. 136

135a Style 1.   **Heath Family, ERICACEAE**

Figure 450

Fig. 450. *Azalea lutea* L., Flame Azalea.

Few flowers make a finer display than the

Azaleas. In April and May our eastern mountains are ablaze with them. Many foreign Azaleas are cultivated.

The Rhododendrons and Mountain-laurel also very showy, belong in this family.

135b Stigma sessile (style wanting or almost so).

Holly Family, AQUIFOLIACEAE

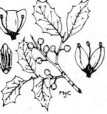

Figure 451

Fig. 451. *Ilex opaca* Ait., American Holly.

Most of the Hollies are evergreen. They are prized for ornamental hedges. There are almost 300 species known, many of which are tropical.

Bees secure much honey from the flowers.

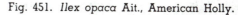

136a Stamens and corolla lobes 4. Figs. 452 and 453. 137

136b Stamens and corolla lobes 5 (or more). Figs. 454 to 458. 138

137a Leaves all arising from the ground, (acaulescent) (a few exceptions), corolla dry and membraneous.

Plantain Family, PLANTAGINACEAE

Figure 452

Fig. 452. a, *Plantago major*, Common Plantain; b, *Plantago arenaria* W. & K., Sand Plantain.

Some 200 species of Plantains are known. Many of them are bad weeds.

Buckhorn or Ribbed Plantain with seeds much the same size as those of red clover is a serious pest.

137b Leaves opposite on stems; corolla normal.

Verbena Family, **VERBENACEAE**

Fig. 453. *Callicarpa americana* L., French Mulberry.

This shrub has pale blue flowers and red-dish-blue fruit. It grows in the southeast.

A number of prized ornamentals and several bad weeds belong in this family of 1300 species.

Figure 453

138a Fruit with 2 or 4 nut-like seeds.

Borage Family, **BORAGINACEAE**

Fig. 454. *Heliotropium peruvianum* L., Common Heliotrope.

This is a favorite house plant on account of its pleasing odor. The flowers are shades of lavender or sometimes white.

Figure 454

138b Fruit a capsule or pod with few to many seeds.
Figs. 455 to 458. 139

139a Twining (or trailing) vines, flowers usually showy.
Morning-glory Family, **CONVOLULACEAE**

Fig. 455. a, *Convolvulus sepium* L., Hedge Bindweed; b, *Ipomoea quamoclit* (L.), Cypress Vine.

Some members of the family are prized for ornamentation; some very serious weeds are also included. The sweet-potatoes valued for food are members of the family.

Figure 455

139b Not twining. Figs. 456 to 458. 140

140a Styles 2; pod many-seeded; leaves alternate, entire.
Water-leaf Family, HYDROPHYLLACEAE

Fig. 456. *Nama ovata* (Nutt.).

It grows in wet soil and is southern in location. It is a showy plant.

The some 45 members of this genus are mostly tropical.

Figure 456

140b Styles but one, though often branched. Figs. 457 and 458. 141

141a Style dividing into 3 linear stigmas.
Phlox Family, POLEMONIACEAE

Fig. 457. a, *Phlox divaricata* L.,

Wild Blue Phlox; b, *Polemonium reptans* L., Greek Valerian; c, *Gilia congesta* Hook., Round-headed Gilia.

Many beautiful garden plants are included here.

Figure 457

141b Style simple ending in a single terminal stigma.
Potato Family, SOLANACEAE

Figure 458

Fig. 458. a, *Physalis alkekengi* L. Chinese Lantern Plant, b, *Solanum nigrum* L., Black Nightshade; c, *Datura stramonium*, Jimson-weed.

Some prominent food plants such as "Irish" Potato, Tomato, Egg Plant, and Peppers belong to the family as does also Tobacco.

142a Fertile stamens only 2. Figs. 459 and 460. 143

142b Fertile stamens 4, in two pairs. Figs. 461 and 462. . . . .**144**
143a Herbs with leaves arising at or near the ground.
Plantain Family PLANTAGINACEAE

Figure 459

Fig. 459. *Plantago elongata* Pursh., Slendeɪ Plantain.

The plantains have ribbed leaves; the flowers are small and whitish. They often grow on poor soil, which is also true of many others of this genus.

143b Trees or shrubs.   Olive Family, OLEACEAE

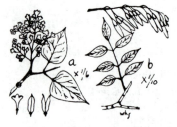

Figure 460

Fig. 460. a, *Syringa vulgaris* L., Lilac; b, *Fraxinus americana* L., White Ash.

Some 30 species of Lilacs, natives of Europe and Asia, are known. They are old favorites for home planting. The flowers are usually of a lavender or purplish shade but white and other colors have been developed.

Figure 461

144a Ovary with 2 cells; each bearing many seeds. Acanthus Family, ACANTHACEAE

Fig. 461. *Ruellia ciliosa* Pursh., Hairy Ruellia.

This plant resembles the common petunia but is not even closely related to it for petunias belong to the family Solanaceae.

144b Ovary with 2 or 4 cells but with only one seed to a cell.
Verbena Family, VERBENACEAE

Figure 462

Fig. 462. *Verbena canadensis* (L.), Large-flowered Verbena.

Here is a fairly large family. In our region only herbs are represented but the family contains many shrubs and trees as found in warmer regions. Some prized flowering plants and several persistent weeds are abundant with us.

145a Fertile stamens (with anthers) 2 or 4. Figs. 464 to 472. 146

145b Fertile stamens 5.   Figwort Family, SCROPHULARIACEAE

Fig. 463. a, *Verbascum thapsus* L., Common Mullen; b, *Verbascum blattaria* L., Moth Mullen.

The mullens are native of the old world. There are well over 100 species. Some of these have been introduced as weeds with us.

Figure 463

146a With but one seed in each cell of ovary or fruit. Figs. 464 to 466. 147

146b More than one seed in each cell of ovary or fruit. Figs. 467 to 472. 149

147a Ovary with 4 lobes and thread-like style arising at intersection of the dividing grooves. Simple opposite leaves; stems mostly square.   Mint Family, LABIATAE

Fig. 464. a, *Nepeta cataria*, Catnip; b, *Monarda fistulosa* L., Horse Mint.

Just why cats should be so vitally concerned with catnip is hard to say, but their behavior in its presence shows that there is some relationship.

Figure 464

147b Ovary without lobes. Figs. 465 and 466. 148

148a Fruit turned downward, with but 1 cell and 1 seed; leaves simple, flowers purplish.   Lopseed Family, PHRYMACEAE

Fig. 465. *Phryma leptostachya* L., Lopseed.

The only species of its family. Fairly common in woods and thickets throughout the eastern U. S.

It is a plant of peculiar appearance and once known may be easily remembered.

Figure 465

151

148b Fruit not turned downward, 2 to 4-cells each with a single seed.
Verbena Family, VERBENACEAE

Fig. 466. *Lippia cuneifolia* Steud., Fog-fruit.

This interesting plant grows in great abundance in low wet ground. The genus has many species in the tropics.

Figure 466

149c Ovary apparently several-celled (actually 1-celled) due to false partitions and diffused placentae.
Unicorn-plant Family, MARTYNIACEAE

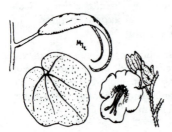

Fig. 467. *Martynia louisiana* Mill., Unicorn-plant.

This plant is sometimes raised as a curiosity but is a native of our midwest and southwest. The flowers are whitish and mottled with purple and yellow.

Figure 467

150a Whitish, yellowish, or purplish parasites on the roots of other plants.
Broomrape Family, ORORANCHACEAE

Fig. 468. a, *Orobanche uniflora* L., Cancer-root; b, *Orobanche fasciculata* Nutt., Yellow Cancer-root.

Parasites are rather unusual among the seed-bearing plants. A few species have so completely acquired the "W. P. A. habit" that they are no longer able to become independent.

Figure 468

150b Not parasitic; equatic plants in which leaves bear minute trap-like bladders, or on moist ground.
Bladderwort Family, LENTIBULARIACEAE

Figure 469

Fig. 469. a, *Utricularia vulgaris* L., Greater Bladderwort; b, *Pinguicula vulgaris* L., Butterwort.

Bladderwort, of which there are several species, grows in shallow water with its conspicuous yellow flowers arising a few inches above the surface. Many tiny bladders serve to trap small aquatic animals.

151a Trees, shrubs or woody vines, leaves opposite; seeds attached to walls of ovary.
Trumpet-creeper Family, BIGNONIACEAE

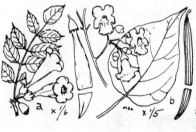

Figure 470

Fig. 470. a, *Tecoma radicans* (L.), Trumpet-creeper; b, *Catalpa speciosa* Warder., Northern Catalpa. The Trumpet-creeper with its large vermilion flowers makes a very showy vine. Catalpa wood is valued for fence posts because of its resistance to decay. The showy purple-flowered Brazilian Jacaranda, much planted in the South, belongs here.

151b Seeds attached to central axis; usually herbs.
Figs. 471 and 472. 152

152a Seeds few, on curved projections.
Acanthus Family, ACANTHACEAE

Figure 471

Fig. 471. *Dianthera americana* L., Dense-flowered Water Willow.

It is common in creeks and other wet places throughout the middle U. S. The flowers are lavender or sometimes whitish.

This is the only species of its genus but the family is large with 2000 known plants.

152b Seeds not borne on curved projections, usually numerous; filaments of stamens usually covered with hairs.

Figwort Family, SCROPHULARIACEAE

Figure 472

Fig. 472. a, *Linaria vulgaris* L. Butter and Eggs; b, *Veronica peregrina* L., Purslane Speedwell; c, *Gerardia purpurea* L., Purple Gerardia; d, *Pedicularis canadensis* L., Lousewort.

Snapdragons, Kenilworth Ivy, and Foxgloves are among the well-known ornamentals of this large family.

A PHYLOGENETIC LIST OF THE FAMILIES OF PLANTS

pinions differ radically among men who are free to think, so it's no wonder that botanists do not always agree on the systematic arrangement of plant families. The list which follows is an attempt to name and arrange the families in logical order for checking purposes.

The list should help in giving an orderly picture of the entire plant kingdom. If the student will check each family here as he learns to recognize it, he will have a chart showing his progress. An approximate estimate of the number of known species is given for some of the groups in parentheses following the group name.

————————*————————

Division THALLOPHYTA
Sub-division Phycophyta (Algae)
Phylum 1 SCHIZOPHYTA
(Blue Green Algae) - (1500)
Class 1 MYXOPHYCEAE
Order 1 Chroococcales
1. Chroococcaceae
2. Entophysalidaceae
Order 2 Chamaesiphonales
1. Pleurocapsaceae
2. Chamaesiphonoceae
Order 3 Hormogonales
1. Oscillatoriaceae
2. Nostocaceae
4. Scytonemataceae
4. Stigonemataceae
5. Rivulariaceae

Phylum 2 EUGLENAPHYCEAE
(350)
Order 1 Euglenales
1. Euglenaceae
2. Colociaceae
3. Astaciaceae
4. Peranemaceae
Phylum 3 CHLOROPHYTA
(Green Algae) (5500)
Class 1 CHLOROPHYCEAE
Order 1 Volvocales
1. Polyblepharidaceae
2. Chlamydomonadaceae

3. Phacotaceae
4. Volvocaceae
5. Spondylomoraceae
6. Sphaerellaceae
Order 2 Tetrasporales
1. Palmellaceae
2. Tetrasporaceae
3. Chlorangiaceae
4. Coccomyxaceae
Order 3 Ulotrichales
1. Ulotrichaceae
2. Microsporaceae
3. Cylindrocapsaceae
4. Chaetophoraceae
5. Protococcaceae
6. Coleochaetaceae
7. Trentepohliaceae
Order 4 Ulvales
1. Ulvaceae
2. Schizomeridaceae
Order 5 Schizogoniales
1. Schizogoniaceae
Order 6 Cladophorales
1. Cladophoraceae
2. Sphaeropleaceae
Order 7 Oedogoniales
1. Oedogoniaceae
Order 8 Zygnematales
1. Zygnemataceae
2. Mesotaeniaceae
3. Desmidiaceae

155

Order 9 Chlorococcales

1. Chlorococcaceae
2. Endosphaeraceae
3. Characiaceae
4. Protosiphonaceae
5. Hydrodictyaceae
6. Coelastraceae
7. Oocystaceae
8. Scenedesmaceae

Order 10 Siphonales

1. Bryopsidaceae
2. Caulerpaceae
3. Halicystaceae
4. Codiaceae
5. Derbesiaceae
6. Vaucheriaceae
7. Phyllosiphonaceae

Order 11 Siphonocladiales

1. Valoniaceae
2. Dasycladaceae

Phylum 4 CHRYSOPHYTA
(6000)

Class 1 HETEROKONTAE

Order 1 Heterochloridales

1. Chloramoebaceae

Order 2 Rhizochloridales

1. Stipitococcaceae

Order 3 Heterocapsales

1. Chlorosaccaceae
2. Mischococcaceae

Order 4 Heterococcales

1. Halosphaeraceae
2. Botryococcaceae
3. Chlorotheciaceae
4. Ophiocytiaceae

Order 5 Heterotrichales

1. Tribonemataceae
2. Monociliaceae

Order 6 Heterosiphonales

1. Botrydiaceae

Class 2 CHRYSOPHYCEAE

Order 1 Chrysomonadales

1. Chromulinaceae
2. Mallomonadaceae
3. Syncryptaceae
4. Synuraceae
5. Ochromonadinaceae
6. Physomonadaceae

Order 2 Rhizochryidales

1. Rhizochrysidaceae

Order 3 Chrysocapsales

1. Chrysocapsaceae
2. Hydruraceae

Order 4 Chrysotrichales

1. Phaeothamniaceae

Class 3 BACILLARIEAE
(Diatoms) (5000)

Order 1 Centrales

1. Coscinodiscaceae
2. Eupodiscaceae
3. Rhizosoleniaceae
4. Chaetoceraceae
5. Biddulphiaceae
6. Anaulaceae

Order 2 Pennales

1. Tabellariaceae
2. Meridionaceae
3. Diatomaceae
4. Fragilariaceae
5. Eunotiaceae
6. Achnanthaceae
7. Naviculaceae
8. Gomphonemataceae
9. Cymbellaceae
10. Nitzschiaceae
11. Surirellaceae

Phylum 5 PYRROPHYTA
(975)

Class 1 DINOPHYCEAE

Order 1 Gymnodiniales
Order 2 Peridiniales
Order 3 Dinophysidales
Order 4 Rhizodiniales
Order 5 Dinocapsales
Order 6 Dinotrichales
Order 7 Dinococcales

Phylum 6 PHAEOPHYTA
(Brown Algae) (1000)

Class 1 ISOGENERATAE
Order 1 Ectocarpales
1. Ectocarpaceae

Order 2 Sphacelariales
1. Sphacelariaceae

Order 3 Tilopteridales
1. Tilopteridaceae

Order 4 Cuteriales
1. Cutleriaceae

Order 5 Dictyotales
1. Dictyotaceae

Class 2 HETEROGENERATAE

Order 1 Chordariales
1. Chordariaceae
2. Leathesiaceae

Order 2 Sporochnales
1. Sporochnaceae

Order 3 Desmarestiales
1. Desmarestiaceae

Order 4 Punctariales
1. Scytosiphonaceae
2. Asperococcaceae
3. Coilodesmaceae
4. Punctariaceae

Order 5 Dictyosiphonales
1. Dictyosiphonoceae

Order 6 Laminariales
    (Kelps)
1. Chordaceae
2. Phyllariaceae
3. Laminariaceae
4. Alariaceae
5. Lessoniaceae

Class 3 CYCLOSPOREAE

Order 1 Fucales
1. Fucaceae
2. Sargassaceae

Phylum 7 RHODOPHYTA
(Red Algae) (2500)

Class 1 RHODOPHYCEAE
Order 1 Bangiales
1. Bangiaceae

Order 2 Nemalionales
1. Chantransiaceae
2. Batrachospermaceae
3. Bonnemaisonaceae
4. Helminthocladiaceae
5. Chaetangiaceae

Order 3 Gelidiales
1. Gelidiaceae

Order 4 Cryptonemiales
1. Dumontiaceae
2. Squamariaceae
3. Corallinaceae
4. Endocladiaceae

Order 5 Gigartinales
1. Cruoriaceae
2. Gracilariaceae
3. Plocamiaceae
4. Rissoellaceae
5. Solieriaceae
6. Rhodophyllidaceae
7. Hypneaceae
8. Gigartinaceae

Order 6 Rhodymeniales
1. Champiaceae
2. Rhodymeniaceae

Order 7 Ceramiales
1. Ceramiaceae
2. Delesseriaceae
3. Rhodomelaceae

Phylum 8 CHAROPHYTA
(200)

Class 1 CHAROPHYCEAE
Order 1 Charales
1. Characeae

Phylum 9 LICHENES
(Lichens) (15,000)

Order 1 Basidiolichenes

Order 2 Ascolichenes
1. Caliciaceae
2. Graphidaceae
3. Lecanactidaceae
4. Gyalectaceae
5. Lecideaceae
6. Psoraceae

7. Baeomycetaceae
8. Cladoniaceae
9. Stereocaulaceae
10. Collemaceae
11. Pyrenopsidaceae
12. Ephebaceae
13. Pannariaceae
14. Stictaceae
15. Peltigeraceae
16. Gyrophoraceae
17. Lecanoraceae
18. Pertusariaceae
19. Parmeliaceae
20. Teloschistaceae
21. Physciaceae
22. Verrucariaceae
23. Pyrenulaceae
24. Dermatocarpaceae
25. Endocarpaceae
26. Leprariaceae

Phylum 10 SCHIZOMYCETES
(Bacteria)

Order 1 Eubacteriales
(True Bacteria)
1. Nitrobacteriaceae
2. Coccaceae
3. Spirillaceae
4. Bacteriaceae
5. Bacillaceae

Order 2 Actinomycetales
1. Actinomycetaceae
2. Mycobacteriaceae

Order 3 Chlamydobacteriales
1. Chlamydobacteriaceae

Order 4 Thiobacteriales
1. Rhodobacteriaceae
2. Beggiatoaceae
3. Achromatiaceae

Order 5 Myxobacteriales
1. Myxobacteriaceae

Order 6 Spirochaetales
1. Spirochaetaceae

Order 7 Caulobacteriales
1. Gallionellaceae

Order 8 Rickettsiales
1. Rickettsiaceae

Phylum 11 MYXOTHALLOPHYTA
(450)

Class 1 MYXOMYCETAE

Order 1 Endosporales
1. Trichiaceae
2. Lycogalaceae
3. Cribrariaceae
4. Stemonitaceae
5. Physaraceae

Order 2 Exosporales
1. Ceratiomyxaceae

Class 2 PHYTOMYXINAE

Order 1 Plasmodiophorales
1. Plasmodiophoraceae
Class 3 ACRASIEAE

Order 1 Acrasiales
1. Acrasiaceae
2. Dictyosteliaceae

Phylum 12 EUMYCETES
(True Fungi) (75,000)

Class 1 PHYCOMYCETES
(Algal-like Fungi) (15,000)

Order 1 Chytridales
1. Rhizidiaceae
2. Olpidiaceae
3. Synchytriaceae
4. Cladachytriaceae
5. Woroninaceae

Order 2 Blastocladiales
1. Blastocladiaceae

Order 3 Monoblepharidales
1. Monoblepharidaceae

Order 4 Ancylistales
1. Ancylistaceae

Order 5 Saprolegniales
1. Saprolegriaceae
2. Leptomitaceae
3. Pythiaceae

Order 6 Peronosporales
1. Peronosporaceae
2. Albuginaceae

Order 7 Mucorales
1. Mucoraceae
2. Mortierellaceae
3. Choanephoraceae
4. Chaetocladiaceae
5. Piptocephalidaceae

Order 8 Entomophthorales
1. Entomophthoraceae
2. Basidiobolaceae

Class 2 ASCOMYCETES
(Sac Fungi) (24,000)

Order 1 Saccharomycetales
1. Saccharomycetaceae
2. Endomycetaceae

Order 2 Aspergilliales
1. Gymnoascaceae
2. Aspergillaceae
3. Onygenaceae
4. Trichcomaceae
5. Myriangiaceae
6. Elaphomycetaceae
7. Terfeziaceae

Order 3 Erysiphales
1. Erysiphaceae
2. Perisporiaceae
3. Microthyriaceae

Order 4 Hysteriales
1. Hypodermataceae
2. Dichaenaceae
3. Ostropaceae
4. Hysteriaceae
5. Acrospermaceae

Order 5 Phacidiales
1. Stictidaceae
2. Tryblidiaceae
3. Phacidiaceae

Order 6 Pezizales
1. Pyronemaceae
2. Pezizaceae
3. Ascobolaceae
4. Helotiaceae
5. Mollisiaceae
6. Celidiaceae

7. Patellariaceae
8. Cenangiaceae
9. Cordieritidaceae
10. Cyttariaceae
11. Caliciaceae

Order 7 Tuberales
1. Tuberaceae

Order 8 Helvellales
1. Geoglossaceae
2. Helvellaceae
3. Rhizinaceae

Order 9 Exoascales
1. Exoascaceae
2. Ascocorticiaceae

Order 10 Hypocreales
1. Hypocreaceae

Order 11 Sphaeriales
1. Chaetomiaceae
2. Sordariaceae
3. Sphaeriaceae
4. Ceratostamataceae
5. Cucubitariaceae
6. Coryneliaceae
7. Amphisphaeriaceae
8. Lophiostomataceae
9. Mycosphaerellaceae
10. Pleosporaceae
11. Massariaceae
12. Gnomoniaceae
13. Clypeosphaeriaceae
14. Valsaceae
15. Melanconidaceae
16. Diatrypaceae
17. Melogrammataceae
18. Xylariaceae

Order 12 Dothidiales
1. Dothidiaceae

Order 13 Laboulbeniales
1. Peyritschiellaceae
2. Laboulbeniaceae
3. Zodiomycetaceae

Class 3 BASIDIOMYCETES
(Club Fungi) (20,000)

Order 1 Ustilaginales
1. Ustilaginaceae
2. Tilletiaceae

Order 2 Uredinales
1. Endophyllaceae
2. Melampsoraceae
3. Pucciniaceae
4. Coleosporaceae

Order 3 Tremellales
1. Auriculariaceae
2. Tremellaceae

Order 4 Hymenomycetales
1. Dacryomycetaceae
2. Exobasidiaceae
3. Hypochnaceae
4. Thelephoraceae
5. Clavariaceae
6. Hydnaceae
7. Polyporaceae
8. Boletaceae
9. Agaricaceae

Order 5 Gasteromycetales
1. Lycoperdaceae
2. Phallaceae
3. Clathraceae

Phylum 13 BRYOPHYTA
(3,000)

Class 1 HEPATICAE (Liverworts)

Order 1 Jungermanniales
1. Ptilidiaceae
2. Lepidoziaceae
3. Calypogeiaceae
4. Cephaloziaceae
5. Cephaloziellaceae
6. Harpanthaceae
7. Jungermanniaceae
8. Marsupellaceae
9. Plagiochilaceae
10. Scapaniaceae
11. Porellaceae
12. Radulaceae
13. Frullaniaceae
14. Lejeuneaceae

Order 2 Metzgeriales
1. Fossombroniaceae
2. Pelliaceae
3. Haplolaenaceae
4. Pallaviciniaceae
5. Metzgeriaceae
6. Aneuraceae

Order 3 Marchantiales
1. Marchantiaceae
2. Rebouliaceae
3. Ricciaceae

Order 4 Sphaerocarpales
1. Sphaerocarpaceae

Order 5 Anthocerotales
1. Anthocerotaceae

Class 2 MUSCI (Mosses)

Order 1 Sphagnales
1. Sphagnaceae

Order 2 Andreaeales
1. Andreaeaceae

Order 3 Bryales
1. Tetraphidaceae
2. Polytrichaceae
3. Fissidentaceae
4. Archidiaceae
5. Ditrichaceae
6. Seligeriaceae
7. Dicranaceae
8. Leucobryaceae
9. Calymperaceae
10. Encalyptaceae
11. Buxbaumiaceae
12. Pottiaceae
13. Grimmiaceae
14. Ephemeraceae
15. Disceliaceae
16. Funariaceae
17. Splachnaceae
18. Schistostegaceae
19. Orthotrichaceae
20. Timmiaceae
21. Aulacomniaceae
22. Bartramiaceae
23. Bryaceae
24. Mniaceae
25. Hypnaceae
26. Leskeaceae
27. Hookeriaceae
28. Neckeraceae
29. Leucodontaceae
30. Cryphaeaceae
31. Fabroniaceae
32. Fontinalaceae

Phylum 14 PTERIDOPHYTA
(Ferns)

Order 1 Filiacles
(4,000)
1. Hymenophyllaceae
2. Polypodiaceae
3. Schizaeaceae
4. Osmundaceae
5. Ophioglossaceae
6. Marsileaceae
7. Salviniaceae
8. Cyatheaceae

Order 2 Equisetales
(30)
1. Equisetaceae

Order 3 Lycopodiales
(500)
1. Lycopodiaceae
2. Selaginellaceae
3. Isoetaceae

Phylum 15 SPERMATOPHYTA
(Seed-bearing Plants)

Class 1 GYMNOSPERMAE (500)

Order 1 Cycadales
1. Cycadaceae

Order 2 Ginkgoales
1. Ginkgoaceae

Order 3 Coniferales
1. Taxaceae
2. Pinaceae

Order 4 Gnetales
1. Gnetaceae

Class 2 ANGIOSPERMAE
(130,000)

Subclass Monocotyledoneae
("Monocots")

Order 1 Pandanales
1. Typhaceae
2. Sparganiaceae

Order 2 Najadales
1. Najadaceae
2. Scheuchzeriaceae

3. Alismaceae
4. Hydrocharitaceae

Order 3 Graminales
1. Gramineae
2. Cyperaceae

Order 4 Palmales
1. Palmaceae

Order 5 Cyclanthales
1. Cyclanthaceae

Order 6 Arales
1. Araceae
2. Lemnaceae

Order 7 Xeridales
1. Eriocaulaceae
2. Xyridaceae
3. Mayacaceae
4. Commelinaceae
5. Bromeliaceae
6. Pontederiaceae

Order 8 Lillales
1. Juncaceae
2. Liliaceae
3. Haemodoraceae
4. Dioscoreaceae
5. Amaryllidaceae
6. Iridaceae

Order 9 Scitaminales
1. Marantaceae
2. Musaceae
3. Cannaceae

Order 10 Orchidales
1. Burmanniaceae
2. Orchidaceae

Subclass Dicotyledoneae
("Dicots")

Order 11 Casuarinales
1. Casuarinaceae

Order 12 Piperales
1. Saururaceae

Order 13 Juglandales
1. Juglandaceae

Order 14 Myricales
  1. Myricaceae

Order 15 Salicales
  1. Salicaceae

Order 16 Fagales
  1. Betulaceae
  2. Fagaceae

Order 17 Urticales
  1. Urticaceae

Order 18 Santalales
  1. Santalaceae
  2. Loranthaceae

Order 19 Proteales
  1. Proteaceae

Order 20 Aristolochiales
  1. Aristolochiaceae

Order 21 Polygonales
  1. Polygonaceae

Order 22 Chenopodiales
  1. Chenopodiaceae
  2. Amaranthaceae
  3. Phytolaccaceae
  4. Nyctaginaceae
  5. Illecebraceae
  6. Aizoaceae
  7. Basellaceae

Order 23 Caryophyllales
  1. Caryophyllaceae
  2. Portulacaceae

Order 24 Ranunculales
  1. Ceratophyllaceae
  2. Nymphaeaceae
  3. Ranunculaceae
  4. Magnoliaceae
  5. Calycanthaceae
  6. Anonaceae
  7. Menispermaceae
  8. Berberidaceae
  9. Lauraceae

Order 25 Papaverales
  1. Papaveraceae
  2. Fumariaceae
  3. Cruciferae

  4. Capparidaceae
  5. Resedaceae

Order 26 Sarraceniales
  1. Sarraceniaceae
  2. Droseraceae

Order 27 Rosales
  1. Podostemaceae
  2. Crassulaceae
  3. Parnassiaceae
  4. Saxifragaceae
  5. Pittosporaceae
  6. Hamamelidaceae
  7. Platanaceae
  8. Rosaceae
  9. Leguminosae

Order 28 Geraniales
  1. Linaceae
  2. Oxalidaceae
  3. Tropaeolaceae
  4. Geraniaceae
  5. Zygophyllaceae
  6. Rutaceae
  7. Simaroubaceae
  8. Meliaceae
  9. Polygalaceae
  10. Euphorbiaceae
  11. Callitrichaceae

Order 29 Sapindales
  1. Buxaceae
  2. Empetraceae
  3. Limnanthaceae
  4. Anacardiaceae
  5. Cyrillaceae
  6. Aquifoliaceae
  7. Celastraceae
  8. Staphyleaceae
  9. Aceraceae
  10. Sapindaceae
  11. Balsaminaceae

Order 30 Rhamnales
  1. Rhamnaceae
  2. Vitaceae

Order 31 Malvales
  1. Tiliaceae
  2. Malvaceae

Order 32 Violales
1. Theaceae
2. Tamaricaceae
3. Hypericaceae
4. Elatinaceae
5. Cistaceae
6. Violaceae
7. Passifloraceae
8. Caricaceae
9. Loasaceae

Order 33 Begoniales
1. Begoniaceae

Order 34 Opuntiales
1. Cactaceae

Order 35 Myrtales
1. Thymelaceae
2. Eleagnaceae
3. Lythraceae
4. Myrtaceae
5. Melastomaceae
6. Onagraceae
7. Haloragidaceae

Order 36 Umbellales
1. Araliaceae
2. Umbelliferae
3. Cornaceae

Order 37 Ericales
1. Ericaceae
2. Diapensiaceae

Order 38 Primulales
1. Plumbaginaceae
2. Primulaceae

Order 39 Ebenales
1. Sapotaceae
2. Ebenaceae
3. Styracaceae

Order 40 Gentianales
1. Oleaceae
2. Loganiaceae
3. Gentianaceae
4. Apocynaceae
5. Asclepiadaceae

Order 41 Polemoniales
1. Cuscutaceae
2. Convolvulaceae
3. Polemoniaceae
4. Hydrophyllaceae
5. Boraginaceae
6. Verbenaceae
7. Labiate
8. Solanaceae
9. Scrophulariaceae
10. Lentibulariaceae
11. Orobanchaceae
12. Bigoniaceae
13. Martyniaceae
14. Gesneriaceae
15. Acanthaceae
16. Phrymaceae

Order 42 Plantaginales
1. Plantaginaceae

Order 43 Rubiales
1. Rubiaceae
2. Caprifoliaceae
3. Valerianaceae
4. Dipsacaceae

Order 44 Campanulales
1. Cucurbitaceae
2. Campanulaceae
4. Compositae
3. Lobeliaceae

PLANTS LIKE DOORS OPEN WITH EASE BY CAREFUL USE OF PICTURED KEYS

INDEX AND PICTURED GLOSSARY

A

Acanthaceae 150, 153
Acanthus Family 150, 153
Acaulescent 147
Acer 125, 131
Aceraceae 125, 131
Acerates 123
Acetobacter 46
Achene, a dry, hard one-seeded indehiscent fruit, 106
Achnanthaceae 25
Acorus 93
Acrasiaceae 49
Acrasiales 49
Acrochaetium 38
Actingstrum 19
Actinomyces 48
Actinomycetaceae 48
Actinomycetales 48
Adder's Tongue 87
Adder's Tongue Family 87
Adiantum 88
Adlumia 139
Aesculus 123
Agar-agar 37
Agaricaceae 67
Ailanthus 127
Ailanthus Family 127
Air Pine 96
Aizoaceae 107
Alaria 36
Alariaceae 36
Albuginaceae 52
Albugo 52
Alder 104
Alfalfa 112
Algae 4, 8
Algal-like Fungi 51
Alisma 94
Alismaceae 94
Alleghany Mountain Spurge 105
Allspice 132
Alnus 104
Alum-root 114
Amaranth Family 109
Amaranthaceae 109
Amaranthus 109
Amaryllidaceae 98
Amaryllis Family 98
Ambrosia 136
Ament, scaly, tassel-like or plume-like inflorescence, 101: (Fig. 473)

Figure 473

American Basswood 116
American Beech 104
American Bladdernut 130
American Brown Rot 58
American Dwarf Mistletoe 103
American Elder 139
American Gray Birch 104
American Heath 117
American Holly 147
American Lotus 110
American Mistletoe 103
American Yew 90
Amoeboid, like an amoeba.
Amphidium 82
Amphora 27
Anabaena 10
Anacardiaceae 121
Ananas 96
Anaulaceae 25
Andreaea 77
Andreaeaceae 77
Andreaeales 77
Anemone 112
Aneuraceae 72
Angiosperm 91
Angiospermae 89
Angle-pod 144
Anise 133
Ankistrodesmus 18
Annulus 88
Anonaceae 111
Anther, the pollen-producing part of a stamen. (Fig. 474)

Figure 474

Antheridium, male reproductive organ 53
Anthoceros 70
Anthocerotaceae 70
Anthocerotales 70
Anthurus 68
Anychia 108
Apiocystis 16
Apiosporium 61
Aplanospore, non-motile asexual spores produced by some algae.
Apocynaceae 144
Apocynum 144
Apothecium, an open cup-like fruiting body, 57: Fig. 475

Figure 475

Apple 113, 134
Apple Scab 63
Apricot 134
Aquifoliaceae 126, 143
Araceae 93
Aralia 134
Araliaceae 134
Arbutus 141
Archegonium, flask-shaped organ bearing an egg in ferns and mosses. (Fig. 476)

LIP CELLS
CANAL CELLS
EGG CELL

Figure 476

Arcyria 51
Arisaema 93
Aristolochia 106
Aristolochiaceae 106
Arrow-grass Family 94
Arrow-head 94
Arrowroot 100
Arrowroot Family 100
Artichoke 136
Arum Family 93
Asci 51
Asclepiadaceae 123, 141
Asclepias 123, 141
Ascobolaceae 58
Ascobolus 58
Ascocarp 57
Ascolichenes 42
Ascomycetes 56
Ascophyllum 36
Aseptate 51
Asexual, non-sexual 33
Ash 131
Asiatic Day-flower 97
Asimina 111
Asparagus 97
Aspergillaceae 61
Aspergillales 61
Aspergillus 61
Asperococcaceae 35
Asplenium 88
Astasia 12
Astasiaceae 12
Aster 136
Asterionella 26
Astrophytum 134
Aulacomniaceae 83
Aulacomnium 83
Auricularia 66
Auriculariaceae 66
Australian Brush-Cherry 132
Australian Pine 100
Authority 3
Autospore 18
Avens 113
Azalea 129, 147

B

Baby's Breath 122
Bacillaceae 47
Bacillarieae 24
Bacillus 47
Bacteria 44
Bacteriaceae 47
Badhamia 50
Balloon-Flower 137
Balsam Family 124
Balsaminaceae 124

INDEX

Bamboo 92
Banana 98
Banana Family 98
Baneberry 112
Bangiaceae 37
Bangiales 37
Banksia 108
Barberry 112, 119
Barberry Family 112, 119
Bartramia 83
Bartramiaceae 83
Basellaceae 112
Basella Family 112
Basidia 51
Basidiobolaceae 55
Basidiobolus '55
Basidiolichenes 41
Basidiomycetes 64
Basswood 116
Batrachospermaceae 38
Batrachospermum 38
Bayberry Family 103
Bean 113
Bed-straw 138
Beech Family 104
Beefwood 100
Beet 109
Begonia 115
Begonia Family 115
Begoniaceae 115
Berberidaceae 112, 119
Berberis 112
Berry, fruit with the flesh
    soft throughout; such as
    tomata, grape, orange.
    (Fig. 477)

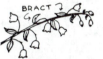

Figure 477

Betula 104
Betulaceae 104
Biddulphia 25
Biddulphiaceae 25
Biennial Gaura 133
Bignoniaceae 153
Birch 104
Birch Family 104
Bird-nest fungus 68
Bird's-foot Violet 122
Birthwort Family 106
Bishop's Cap 134
Bittersweet 125
Black Gum 132
Black Molds 54
Black Mosses 77
Black Nightshade 149
Black Stem Rust of Wheat
    65, 112
Black Walnut 104
Blackberry 113
Black-capped Helvella 59
Black-knot of Cherry 62
Black-rot of Apple and other
    Pome fruits 63
Black-rot of Grapes 63
Bladdernut 130
Bladdernut Family 130
Bladderwort 153
Bladderwort Family 153
Blasia 71
Blazing Star 136
Bleeding Heart 118

Bloodroot 115
Bloodwort Family 99
Blue-bell Family 137
Blue-bells 137
Blueberry 141
Blue-eyed Grass 99
Blue-grass 92
Blue-green algae 8
Bluets 138
Boletaceae 67
Boletus 67
Borage Family 144, 148
Boraginaceae 144, 148
Botrychium 87
Botrydiaceae 30
Botrydiopsis 30
Botrydium 18, 30
Botryococcaceae 30
Botryococcus 30
Bouncing Bet 122
Boussingaultia 112
Box Family 105
Brachiomonas 15
Brachythecium 85
Bract, a leaf-like part sub-
    tending a flower or inflor-
    escence. (Fig. 478)

Figure 478

Brassica 119
Brazilian Jacaranda 153
Broad-fruited Bur-reed 94
Broad-leaved Arrow-head 94
Broad-leaved Cat-tail 93
Brome-grass 92
Bromeliaceae 96
Bromus 92
Broomrape Family 152
Brown Algae 24, 35
Brownian movement 47
Bryaceae 83
Bryales 77
Bryophyta 7, 70
Bryopsidaceae 13
Bryopsis 13
Bryum 83
Buckbrush 139
Buckeye 123
Buckhorn 147
Buckthorn 120
Buckthorn Bumelia 143
Buckthorn Family 120
Buckwheat 106
Buckwheat Family 106
Budding Yeast 56
Buellia 43
Buginvillaea 108
Bulbochaete 20
Bumelia 143
Bundles, vascular; a collec-
    tion of heavy-walled elon-
    gated cells for transport-
    ing liquids in plants.
Bur-reed Family 94
Butter and Eggs 154
Buttercup 112
Butterfly-weed 144
Butterwort 153
Buxaceae 105
Buxbaumiaceae 77

C

Cabbage 119
Cabomba 110
Cactaceae 134
Cactus Family 134
Calamus 93
Caliciaceae 42
Calicium 42
Calla Lily 93
Callicarpa 148
Callithamnion 41
Callitrichaceae 101
Callitriche 101
Caltrop Family 128
Calvatia 68
Calycanthaceae 110
Calycanthus 110
Calycanthus Family 110
Calyptra, old archigonial wall
    covering the capsule of
    mosses. (Fig. 479)

Figure 479

Calyx 100
Campanula 137
Campanulaceae 137
Canada Moonseed 111
Canada Plum 113
Cancer-root 152
Canna 100
Canna Family 100
Cannaceae 100
Canterbury Bells 137
Caper Family 118
Capillitium 50
Capparidaceae 118
Caprifoliaceae 139
Capsella bursa-pastoris 119
Capsule, dry self-opening
    fruit with more than one
    carpel. (Fig. 480)

Figure 480

Caraway 133
Cardamine 130
Cardinal-flower 137
Carex 92
Carica 140
Caricaceae 140
Carnation 122
Carnegiea 132, 134
Carolina Buckthorn 120
Carolina Silverbell 142
Carolina Yellow-eyed Grass
    95

Carpel, a simple pistil or **one** part of a compound pistil. (Fig. 481)

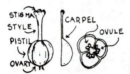

Figure 481

Carpelate Flower, female flower with pistil (carpels) but no stamens. (Fig. 482)

Figure 482

Carpet-weed Family 107
Carpogonium, organ of Red Algae in which carpospores develop.
Carpomitra 34
Carrot 133
Carrot Family 133
Carya 104
Caryophyllaceae 122, 130
Cashew Family 121
Cassia 113
Castalia 110
Castor-oil Plant 127
Casuaria 100
Casuarina Family 100
Casuarinaceae 100
Catalpa 153
Catchfly 130
Catkin, a tassel or plumelike inflorescence; a n ament. (Fig. 483)

Figure 483

Catnip 151
Cat-tail Family 93
Cauliflower 119
Caulobacteriales 45
Ceanothus 120
Celastraceae 125
Celastrus 125
Celery 133
Central Placenta 120

Centrales 24
Cephalozia 74
Cephaloziaceae 74
Ceramiaceae 41
Ceramiales 40
Cerastium 122
Ceratiomyxa 49
Ceratiomyxaceae 49
Ceratium 29
Ceratodon 80
Ceratoneis 25
Ceratophyllaceae 101
Ceratophyllum 101
Chaetangiaceae 39
Chaetophoraceae 22
Chaetosphaeridium 22
Chamaesiphon 9
Chamaesiphonaceae 9
Chamaesiphonales 9
Chantransiaceae 38
Chara 11
Characeae 11
Characiaceae 17
Characiopsis 30
Characium 17
Charales 11
Chenopodiaceae 109
Chenopodium 109
Cherry 113, 134
Chicory 136
Chiloscyphus 74
Chinese Honeysuckle 139
Chinese Lantern Plant 149
Chionanthus 131
Chlamydobacteriaceae 46
Chlamydobacteriales 46
Chlamydomonadaceae 15
Chlamydomonas 15
Chlamydospores 64
Chlorangiaceae 16
Chlorella 18
Chlorobotrys 30
Chlorococcaceae 18
Chlorococcales 15, 17
Chlorococcum 18, 42
Chlorophyceae 12
Chloroplast, small chlorophyll (green) containing body within a cell.
Chlorotheciaceae 30
Chonhrus 40
Chordariaceae 34
Chromatophore, large plant plastid, green or other colors.
Chroococcaceae 9
Chroococcales 8
Chroococcus 9
Chrysanthemum 136
Chrysocapsales 31
Chrysomonadales 31
Chrysophyceae 31
Chytridales 52
Cistaceae 117, 122
Cladochytriaceae 53
Cladochytrium 53
Cladonia 43
Cladoniaceae 43
Cladophora 23
Cladophoraceae 23
Cladophorales 22
Clammy-weed 118
Clathraceae 68
Clavaria 67
Clavariaceae 67
Claviceps 62
Claytonia 119
Clearweed 105
Clematis 112
Cleome 118

Climbing False Buckwheat 106
Climbing Fern 87
Climbing Fern Family 8**7**
Climbing Fumitory 139
Closed Gentian 145
Closterium 23
Clostridium 47
Clover 113
Cloves 132
Club Fungi 51, 64
Club Mosses 87
Club Moss Family 89
Club Rush 92
Club-root of cabbage 49
Coccaceae 47
Coccomyxa 16
Coccomyxaceae 16
Codiaceae 13
Codium 13
Coelastraceae 19
Coelastrum 19
Coenobic cells, the number of cells remaining the same throughout the life of the plant.
Coenocyte, having several nuclei within one cell.
Coffee 138
Colaciaceae 11
Colacium 11
Coleochaetaceae 22
Coleochaete 22
Columella 50
Comandra 105
Comatricha 50
Commelina 97
Commelinaceae 97
Common Amaryllis 98
Common Box 105
Common Dandelion 136
Common Evening-Primrose 136
Common Fern Family 88
Common Flowering Canna 100
Common Garden Currant 114
Common Heliotrope 148
Common Hoptree 126
Common Juniper 90
Common Mullen 151
Common Plantain 147
Common Poke 106
Common Prickly-ash 126
Common Ragweed 136
Common Teasel 138
Common Urn-Moss 81
Compositae 136
Composite Family 136
Composite Flower (Fig. 484)

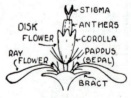

Figure 484

Compound Leaf, one divided into several **leaflets**.
Conceptacle, a cavity containing reproductive cells.

INDEX

Conidia, asexual cells, bud-
ded-off from the end of
an elongated cell, the
conidiophore.
(Fig. 485)

Figure 485

Conidiophore, cell produc-
ing conidia. (Fig. 485)
Conjugation 27
Convolvulaceae 148
Convolvulus 148
Coral Fungi 67
Coral-berry 139
Corallina 39
Cord Moss 81
Coriander 133
Corn 92
Corn Poppy 115
Corn Smut 65
Cornaceae 132
Cornus 132
Corolla 100
Corydalis 118
Corynebacterium 48
Coscinodiscaceae 24
Costaria 36
Cotton 116
Cotyledons, seed-leaves, con-
taining stored food.
(Fig. 486)

Figure 486

Cow Lily 110
Crab Apple 134
Crape-Myrtle 129
Crassulaceae 113
Crataegus 133
Crenothrix 46
Cress 130
Cribraria 50
Cribrariaceae 50
Cronartium 65
Croton 127
Crowfoot Family 112
Cruciferae 119, 130
Crucigenia 19
Crustose lichens 42
Cryptonemiales 39
Ctenocladus 22
Cuba Royal Palm 91
Cucumber 131, 137
Cucumis 137
Cucurbitaceae 131, 137
Cumin 133
Curly-grass 87
Currant 114, 135
Cuscuta 146

Cuscutaceae 146
Custard Apple Family 111
Cutleria 33
Cutleriaceae 33
Cyatheaceae 86
Cycadaceae 90
Cycas 90
Cycas Family 90
Cyclotella 24
Cylindric Blazing Star 136
Cylindrocapsa 21
Cylindrocapsaceae 21
Cylindrocystis 23
Cylindrospermum 10
Cymathaere 36
Cymbellaceae 27
Cynophallus 68
Cyperaceae 92
Cyperus 92
Cypress Vine 148
Cypripedium 99
Cyrilla 126
Cyrilla Family 126
Cyrillaceae 126
Cystoclonium 40
Cythus 68

D

Daffodil 98
Dahlia 136
Daisy 136
Daldinia 64
Dandelion 136
Dayse 41
Datura 149
Day-flower 97
Delesseria 40
Dellesseriaceae 40
Delphinium 112
Dendroid, taking the form
of a tree.
Dendropogon 96
Dense-flowered Water Wil-
low 153
Dermatocarpaceae 42
Dermatocarpon 42
Dermocentroxenus 44
Desmarestia 34
Desmarestiaceae 34
Desmarestiales 34
Desmidiaceae 23
Desmidium 23
Deutzias 114
Dewberry 134
Dianophysis 28
Dianthera 153
Diapensia 146
Diapensia Family 146
Diapensiaceae 146
Diatom 24
Diatoma 26
Diatomaceae 26
Diatomaceous earth 26
Dicentra 118
Dichotomosiphon 13
Dicksonia 86
Dicotyledons 91, 100
Dicranaceae 80
Dicranum 80
Dictosiphonales 35
Dictyosiphon 35
Dictyosiphonaceae 35
Dictyostelium 49
Dictyota 33
Dictyotaceae 33
Diderma 50
Dill 133
Dinobryon 31
Dinococcales 28

Dinophyceae 28
Dinophysidales 28
Dioecious 97
Dioscorea 98
Dioscoreaceae 98
Diospyros 141
Diplococcus 47
Diploid, having the full
number of chromosomes.
Diplolepidious, peristome
teeth with two rows of
scales on the outside
and one row on their
inner side. (See Fig.
231)
Diploneis 27
Dipsacaceae 138
Dipsacus 138
Diptheria 48
Dirca 107
Discoid, disk-shaped.
Ditrichaceae 80
Ditrichum 80
Dock 106
Dodder Family 146
Dodocatheon 143
Dogbane Family 144
Dogwood Family 132
Dothidiaceae 62
Dothidiales 62
Dotted Thorn 133
Downy Brome-grass 92
Downy-mildew 51
Downy-mildew of Grapes 51
Draparnaldia 22
Drosera 121
Droseraceae 121
Drupe, fleshy fruit with
single stone seed as plum
or cherry. (Fig. 487)

Figure 487

Duckweed 91
Duckweed Family 91
Dudresnaya 39
Dumontiaceae 39
Dutchman's Breeches 118
Dwarf Larkspur 112
Dwarf Palmetto 91
Dwarf St. John's-wort 131

E

Ear Fungi 66
Earth Tongues 59
Earthstar 68
Eastern Wahoo 125
Ebenaceae 141
Ebony 141
Ebony Family 141
Echinocactus 134
Ectocarpaceae 32
Ectocarpus 32
Edible Gyromitra 59
Eel Grass 97
Egg Plant 149
Egregia 36
Elater, sterile twisted threads
within a sporangium.
Elaeagnus 107
Elaegnaceae 107

INDEX

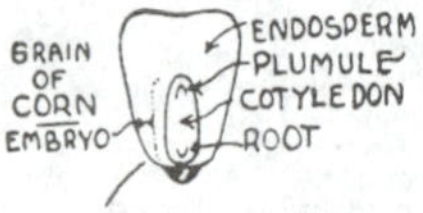

Figure 488

F

Figure 489

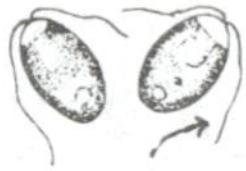

Figure 490

G

INDEX

Haploid, having the half
  number of chromosomes.

Holdfast, the cell or part
  by which an algal plant
  is attached to its support.
  (Fig. 491)

Figure 491

Hymenial layer 58
Hymenium, spore-bearing
  layer in the fungi, 66
Hymenomycetales 66
Hypanthium, special ring-
  like part in some flowers
  to support the stamens, as
  in the rose.
Hypha, one thread-like piece
  of the tissue of a fungus
  plant.

I

Inferior Ovary, arising below
  the calyx, 95
  (Fig. 492)

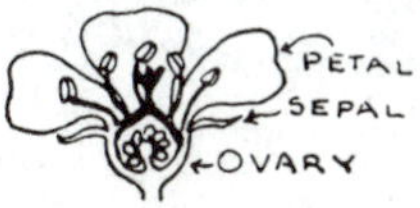

Figure 492

Involucre, circle of bracts
  surrounding a flower or a
  head 69: (Fig. 493)

Figure 493

Isodiametric with all dimen-
  sions approximately equal.
Isogametes, all gametes the
  same size.

J

K

INDEX

L

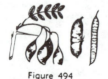

Figure 494

M

Figure 495

Nodules, small fleshy bodies on roots or other parts of plants.

O

Ostiole, the opening in a fruiting body.
Ovary, part of a pistil that contains the ovules (seeds) 89: (Fig. 496)

Figure 496

Ovule, body in the ovary which becomes a seed. (Fig. 496)

P

Palmate Leaf, compound leaf with all the leaflets arising from the end of the petiole. (Fig. 497)

Figure 497

Palmately veined, all the principal veins of a leaf arising at the base of the blade. (Fig. 498)

Figure 498

Panicle, a loose inflorescence with the pedicels branched. (Fig. 499)

Figure 499

Papillae, small projecting parts.
Papillose 78
Paraphyses, sterile cells or threads among fruiting bodies.
Parietal placenta 120

INDEX

Figure 500

Figure 501

Figure 502

Figure 503

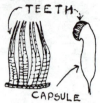

TEETH
CAPSULE

Figure 504

Figure 505

PETAL
SEPAL

Figure 506

LEAFLET
PETIOLE

Figure 507

Figure 508

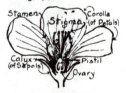

Stamen
Stigma
Corolla (of Petals)
Calyx (of Sepals)
Pistil
Ovary

Figure 509

INDEX

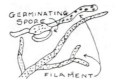

Figure 510

Protoplast, a living cell.
Protosiphon 18
Protosiphonaceae 18
Prunus 113
Psedera 120
Pseudocillium, thread-like
    structure which does not
    move.
Pseudomonas 47
Pseudopodium 77
Pseudoraphe, a longitudinal
    clear space (not cleft) in
    the valve of some diatoms.
Ptelea 126
Pteridophyta 7, 86
Pteromonas 14
Pterospora 141
Ptilidiaceae 73
Ptilidium 73
Puccinia 65
Pucciniaceae 65
Puccoon 144
Puffball 68
Pumpkin 131
Punctariales 35
Purple Cress 130
Purple Gerardia 154
Purple Milkwort 124
Purple-stemmed Cliff-brake
    88
Purslane 135
Purslane Family 115, 119,
    135
Purslane Speedwell 154
Pycnidium, fungus fruiting
    body.
Pycnophycus 37
Pyramimonas 14
Pyrenoid, starch-producing
    body within the chroma-
    tophore of some algae.
Pyrola 129
Pyronema 58
Pyronemaceae 58
Pyrularia 105 .
Pythiaceae 53
Pythium 53

Q

R

Figure 511

Figure 512

Rhizome, an under-ground
    stem for storing food, (a);
    or growing new plants,
    (b). (Fig. 513)

Figure 513

Sepal, one of the leaf-like
parts composing the calyx.
(Fig. 514)

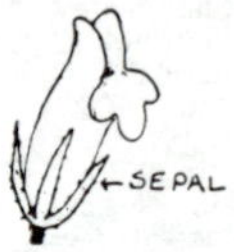

Figure 514

Septa 26
Septate 51
Septum, cross-wall between
plant cells.
Seven-angled Pipewort 96
Shagbark Hickory 104
Sheath, an enclosing part as
the lower part of a grass
leaf. (Fig. 515)

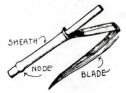

Figure 515

Silicle, a short silique.
(See Fig. 516b)
Silique, fruit of the Mus-
tards; a pod with thin
central partition and opening at both edges. (See
Fig. 516a)

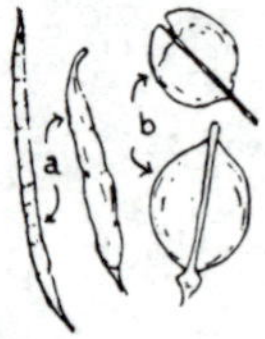

Figure 516

INDEX

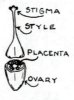
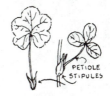

INDEX

INDEX

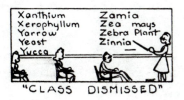

Figure 527

X

Y

Z

Figure 528

"CLASS DISMISSED"